Science and Objectivity

EPISODES IN THE HISTORY OF ASTRONOMY

SCIENCE
AND
OBJECTIVITY

Episodes in the History of Astronomy

Norriss S. Hetherington

Iowa State University Press, Ames

NORRISS S. HETHERINGTON is active in several fields of intellectual endeavor, including inquiry into the nature of science. He has graduate training and research and teaching experience in astronomy, in history, and in the history of science. He is a research associate at the University of California and recent holder of an American Historical Association fellowship.

© 1988 Iowa State University Press, Ames, Iowa 50010
All rights reserved

Composed and printed in the United States of America

FIRST EDITION, 1988

Library of Congress Cataloging-in-Publication Data

Hetherington, Norriss S., 1942–
    Science and objectivity.

    Bibliography: p.
    Includes index.
    1. Astronomy — History.   2. Science — Philosophy.   3. Objectivity.   I. Title.

QB32.H48   1988   520'.9   87–3140
ISBN 0–8138–1159–7

FOR EDITH

Contents

Preface

THE HOLY GRAIL represented man's search for an ideal and for a true religion. The Holy Grail of the twentieth century is objectivity, attained through science. Objectivity is an ideal, no more attainable than the Holy Grail. Nor is objectivity a realized achievement. Nonetheless, science, the search for objectivity, has become the religion of the twentieth century, the standard against which other ideologies are measured.

No doubt later generations will recognize our intellectual foibles as easily as we recognize those of earlier generations — and be is blind to theirs as we are to ours. The Roman historian Tacitus neld up as examples good men and bad, distorting history systematically to do so. Reading Tacitus, the late philosopher R. G. Collingwood was reminded of Socrates' laughter at Glaucon's imaginary portraits of perfectly good and perfectly bad men: "My word, Glaucon, how energetically you are polishing them up like statues for a prize competition."[1] Now we polish up images of scientists, the good objectively determining truth, the bad led astray by subjective values.

Scientists know these polished images to be outrages upon history, but it is not their profession to correct bad histories. Nor might it be prudent. Heretics, dissenting from accepted belief or established dogma, have not been dealt with kindly in earlier epochs. Furthermore, the polished image of science may help dazzle potential patrons.

In contemporary, enlightened times, the worst fate of heretics is to be ignored. So the English astronomer Hermann Bondi learned when he submitted to the Royal Astronomical Society a paper reciting past mistakes of astronomers and concluding that theory often had proven more reliable than observation. A majority of the Council of the Society wished to reject the paper. Fred

Hoyle, then on the Council and in this instance in the minority, noted that all of Bondi's examples were taken from well-attested literature. "Was a paper to be rejected because its statements were correct?" Hoyle asked. Apparently not. In an anguished voice, the president of the society plaintively pleaded, "Then will somebody propose that this paper be rejected irrespective of its contents."[2] Bondi's paper was not published by the Royal Astronomical Society. It appeared later, under less prestigious auspices, to less notice than it might otherwise have received.[3]

Occasionally small obstacles have been raised against historical studies potentially threatening the orthodox image of science. Extensive and controversial scientific material from the papers of the American astronomer Edwin Hubble form part of the critique of the supposed objectivity of science—notwithstanding an initial refusal to open the papers for study on the grounds that only personal matters were contained therein. And there was talk at one time of banning scholars from the archive of the world's greatest observatory, lest disrespectful historians find there grist for their critical mills.[4] The concern was not totally misplaced, even if historians possessed already an abundance of material. No need to release additional skeletons from closed closets. Also, it may have been some of the same scientists intent on barring the doors of their archive who were soon to view as regrettable a historical note on an early study of Mars, the note's publication unfortuitously coinciding with a request for financial support for new Martian studies.[5] But generally, the response to critical studies of science and the quest for objectivity has not been unfriendly, just unenthusiastic or uninterested.

Surprisingly little effort has been expended upon an examination of scientific objectivity, given its central place in modern civilization. In his book *Science and Human Values,* Jacob Bronowski observed that "the sanction of experienced fact as a face of truth is a profound subject, and the mainspring which has moved our civilization since the Renaissance."[6] Bronowski did not exaggerate the importance of science, but he did accept without question its presumed objectivity. That presumption dictated his subsequent strategy for refuting the suggestion that science and values might belong to different worlds, that the world of *what* is subject to test while the world of *what ought to be* is not. Attempting to bring together science and values, he argued that the world of values is subject to test, as is science. He might better have examined the sanctity of scientific fact and found it not completely objective, not completely different from human values.

A thorough test of our intellectual foundation remains to be done. Occasional cracks in the foundation have appeared but have yet to elicit much concern.

Some lack of concern is to be expected among those who know science and know that objectivity is more ideal than achievement. To professional scientists and historians of science, a critique of supposed objectivity in science may appear scarcely more productive than the flogging of a dead horse. Indeed, such was the response of a reader, who nonetheless recommended publication of this study of science and objectivity. Why flog further a thesis well known? Because beyond a small circle the objectivity of science is widely assumed and little questioned. This horse is alive for many, yet to be let out of the stable.

The examples of failed objectivity in science produced and examined here are taken, without exception, from the history of astronomy. It is not that astronomers are in possession of a natural monopoly. At work is a selection effect; quite simply, these are the examples known to the author. Historians of other sciences no doubt can produce examples from their own fields with sufficient understanding and in sufficient detail to convince even a devout congregation that objectivity in science has failed more than once.

The demonstration of failed objectivity is relatively straightforward. The more difficult task ahead is to add to an understanding of how and why objectivity has failed. To add to an understanding of science and the quest for objectivity is a daunting task.

Also daunting is the author's traditional duty to thank those whose help has brought a book into being. Footnotes acknowledge the studies of others upon which this book is built. There will be found the names of pioneers in the history of astronomy, such as R. H. Austin, Richard Baum, J. A. Bennett, Richard Berendzen, Terri Bloom, Stephen Brush, Michael Crowe, Stillman Drake, Don

Fernie, Owen Gingerich, Morton Grosser, Richard Hart, Michael Hoskin, William Hoyt, Stanley Jaki, Kenneth Jones, W. H. McCrea, John North, Donald Osterbrock, Colin Ronan, Simon Schaffer, Robert Smith, Debbie Warner, G. J. Whitrow, and Mari Williams.

Financial assistance came from the National Science Foundation, the National Endowment for the Humanities, the National Aeronautics and Space Administration, the Huntington Library, the Dudley Observatory, York University, and the University of Kansas. Other institutions opening their doors to the author were the Cambridge, Dunlap, Hale, Harvard, Leiden, Lick, and Lowell observatories and the universities of California, Cambridge, Indiana, and Oklahoma.

Suzanne C. Lowitt sympathetically edited the manuscript.

The most important acknowledgment is to colleagues, friends, and family, who fanned the research fire when it burnt low. Sydney van den Berg, Tom Cowling, David Dewhirst, Clark Elliott, Ivan King, David Lindberg, Ray Lyttleton, E. Öpik, and Robert Rosenthal took an interest in the problem of objectivity and directed my attention to important instances, issues, and sources. Steve Brush and David Edge also generously supplied encouragement when it was most needed, and Gibson Reaves as editor of the Astronomical Society of the Pacific leaflet series chose to publish my earliest writings. Mike Crowe read an early draft of this book; he offered suggestions absolutely vital to the book; and his unfailing encouragement saw the book through to completion. Above all, my wife Edith supported the project.

Science and Objectivity

EPISODES IN THE HISTORY OF ASTRONOMY

1

Images of Science

IN THE AFTERMATH and ruin of World War I the ideal, or idol, of objective science was an idea whose time had come. Western civilization's material culture had been rent by the Great War; its intellectual culture had been shocked by the brutal nature of the war and by the fact of war itself. Immediately touched was Henry James's prescient sensitivity. On 5 August 1914 he wrote that "the plunge of civilization into this abyss of blood and darkness . . . is a thing that so gives away the whole long age during which we have supposed the world to be, with whatever abatement, gradually bettering, that to have to take it all now for what the treacherous years were all the while really making for and *meaning* is too tragic for any words."[1] After the war George Bernard Shaw in the play *Too True to be Good* wrote of a civilization led not to the millenium but to suicide, of government thugs left running about with a great burden of corpses and debt, and asking what they must do to be saved.[2]

Salvation was sought in science, although — paradoxically — changes in science had played a part second only to that of the Great War in the destruction of European civilization. Darwinian evolution had begun the weakening of civilization's intellectual underpinnings before the war, and Einsteinian relativity contributed to the feeling of lost directions and values after the war. Before Darwin, noted Shaw, consolation and reassurance were to be found in natural history museums in the contemplation of Purpose and Design. But now Darwin's operation of natural selection had replaced the agency of any designer. Furthermore, Newton's rational Determinism, an impregnable citadel of modern civilization for three hundred years, had crumbled like the walls of Jericho before the criticism of Einstein. Shaw's hero was in despair; he had discovered that the only trustworthy dogma was that there is no dogma.

Undoubtedly the iconoclast Shaw was not unhappy at the prospective demise of all dogma. Lesser mortals, however, felt the loss and sought a substitute. Despite the role of science in the destruction of previous values—or perhaps all the more so because of the power of science—it was to science that much of mankind now turned in search of a new dogma. Objective science was a necessary totem for a civilization that had lost faith in its other conventions.

One exuberant and ultimately influential spokesman for science was George Sarton. He was born in 1884, the son of a Belgian railway engineer. Sarton began philosophical studies at the University of Ghent, only to abandon them in disgust and switch to the natural sciences. He graduated in 1911 determined to devote his life to the disinterested study of mathematical methodology and the history of science, endeavors then offering no prospect of financial support for Sarton and his bride. Failing to obtain a sinecure in the state bureaucracy, Sarton sold his deceased father's wine cellar and with the proceeds purchased a house and founded a journal for the history of science. The temporality of aspects of civilization other than science was brought home to Sarton two years later when he found it necessary to bury his notes in the garden and flee Belgium before the advancing Huns.[3] Old values had proven insufficient to check the worst in man. Science was the only survivor of the war. It alone among human activities could claim to be truly cumulative and progressive.

Science alone of human activities had lasting worth. Sarton's later praise for science was even phrased in the language of science, in the form of classical geometry:

Definition: Science is systematized positive knowledge, or what has been taken as such at different ages and in different places.

Theorem: The acquisition and systematization of positive knowledge are the only human activities which are truly cumulative and progressive.

Corollary: The history of science is the only history which can illustrate the progress of science.[4]

For Sarton, science was the solid foundation upon which to build a lasting and progressive civilization.[5]

Among those influenced by and sharing Sarton's vision of cumulative, objective science was the American astronomer Edwin Hubble.[6] He furnishes another illustration of the idolatry of objective science in the years after World War I. To Sarton's philosophy, Hubble added a second contemporary definition of science; science consisted of judgments upon which universal agreement can be obtained.[7] Hubble also mixed in ideas from his legal training (he had read Roman law as a Rhodes scholar at Oxford and then practiced briefly in the United States). Reminders of law are especially evident in the choice of language and images and also are to be found in Hubble's skillful orchestration of scientific evidence.[8] Hubble found modern science a major advance, differing from medieval science as does experiment from dialectics.

Science deals with judgments on which it is possible to obtain universal agreement. These judgments do not concern individual facts and events, but the invariable association of facts and events known as the laws of science. Agreement is secured by observation and experiment — impartial courts of appeal to which all men must submit if they wish to survive. The laws are grouped and explained by theories of ever increasing generality. The theories at first are ex post facto — merely plausible interpretations of existing bodies of data. However, they frequently lead to predictions that can be tested by experiments and observations in new fields, and, if the interpretations are verified, the theories are accepted as working hypotheses until they prove untenable. The essential requirements are agreement on the subject matter and the verification of predictions. These features insure a body of positive knowledge that can be transmitted from person to person, and that accumulates from generation to generation.

It should be emphasized that the necessity for universal agreement completely bars science from the great world of values, the world in which we spend most of our lives. Judgments of beauty and reverence, for instance, are purely personal judgments. These are not external tests of validity; there is no calculus of values. Wisdom cannot be directly transmitted, and does not readily accumulate through the ages.[9]

It was these considerations, the possibility of universal agreement and of positive, cumulative knowledge, that separated for Hubble the world of science from the world of values. A scientist "naturally and inevitably mulls over the data and guesses at a solution." He proceeds to "testing of the guess by new data—predicting the consequences of the guess and then dispassionately inquiring whether or not the predictions are verified."[10]

Hubble did not always practice precisely what he preached. Indeed, in selecting a model of the universe, he rejected the decision handed down by the court of observation and obeyed instead the judgment from the greater world of values.[11] But the image of objective science shone on as a beacon lighting man's way to true progress.

A particularly elegant development and summing up of the idea of objective science is found in the work of the historian C. C. Gillispie. The central theme of *The Edge of Objectivity,* as the title indicates, is objective science. According to Gillispie, modern science in contrast to Greek science takes its starting point outside the mind; objective observation is its base. Science, objective science, is the distinctive achievement of our history, independent of transitory values. The history of science since the sixteenth century is the history of the advancing edge of objectivity.[12]

The image of objective science is attractive and assuring. It alone stands as guarantor of human progress. But the picture of objective science and civilization is flawed; the picture is not a perfectly accurate representation of reality. Fiction here differs from fact.

Observation statements are not always universally independent of scientists' expectations, are not always objective, as accumulating evidence increasingly demonstrates. Even in the field of twentieth-century physics, the touchstone of modern science, there are instances in which a purported experimental effect has later turned out not to exist. To the innumerable cases in which different groups of physicists measuring the same parameter have reported results differing by more than the claimed limits of error may be added a few more serious instances in which foreseen, anticipated, even wanted results were found by scientists adhering to generally accepted scientific methods—even though the reported phenomena do not exist.[13]

Nor is there blind, universal faith among physicists in experimental data, in objective as opposed to subjective factors. According to Nobel-laureate Murray Gell-Mann: "When you have something simple that agrees with all the rest of physics and really seems to explain what's going on, a few experimental data against it are no objection whatsoever."[14]

A similar evaluation of the relative reliability of theory and observation is found in astronomy, along with several examples of the failure of objectivity. Writing thirty years ago, the English astronomer Hermann Bondi expressed his sense that in astronomy observations had proven a less reliable arbiter of hypotheses than had theoretical considerations.[15] A few cases were then known and were cited by Bondi in support of his contention. Recent historical studies have widened and deepened knowledge of some of the cases cited by Bondi and have uncovered additional instances in the history of astronomy in which, notwithstanding the rigors of the scientific method, attempts to stem a flood of personal biases into the measurement of empirical fact were unavailing.[16]

Not all, nor even a majority of such instances are relegated to ancient history. On the sheer basis of numbers, more instances of failed objectivity must be expected to have occurred during the current century than in earlier times. Roughly seven of every eight scientists who ever lived are alive today; most of the science produced has been produced within living memory.[17]

The changing nature of science also suggests an increase in instances of failed objectivity. Science never completely fit the naive Baconian inductive model in which all facts were collected and then inevitable theories were inevitably induced. That model flounders upon an infinity of possible observations; true Baconian science would never advance beyond the infinite period of fact gathering to the stage of theory formulation. Modern science increasingly deviates from the Baconian inductive model; theories increasingly are used to suggest observations of potential significance.[18] In scientific discovery, "the catalytic role of intuition and hypothesis is essential in making sense of disjoint empirical results and in mapping out the search for new data."[19]

The employment of hypotheses or expectations, essential to scientific progress, also makes inevitable failures of objectivity. The

American astronomer Henry Norris Russell warned: "if the observer knows in advance what to expect . . . his judgment of the facts before his eyes will be warped by this knowledge, no matter how faithfully he may try to clear his mind of all prejudice. The preconceived opinion unconsciously, whether he will or not, influences the very report of his senses, and to secure trustworthy observations, it has been recognized everywhere, and for many years, he must keep himself in ignorance of what he might expect to see."[20]

It is now recognized that scientists cannot keep themselves in complete ignorance, oblivious to all possibilities. Inherent in modern science, increasingly with the advance of science, is the risk of failed objectivity.

Notwithstanding the implausibility of ever fully achieving completely objective science, human nature being what it is, the image of objective science persists, even in the presence of well-attested evidence of failures of objective science. It persists because a psychological need for it persists. It also persists because the image of objective science itself throws up a barrier against a deeper understanding of science. A historian believing in objective science is unlikely to search diligently for counterinstances. A sociologist believing in objective science, believing that scientific representations are determined by the nature of reality, is unlikely to attempt a sociological account of the production and evaluation of scientific knowledge.[21]

Doubts about, even objections to the applicability of historical and sociological analysis to scientific knowledge and practice arise from a sense that scientific knowledge is strictly controlled by observation statements that in turn are established beyond reasonable question by rigorous scientific method. Preconceptions or social conditions may encourage or discourage a scientist in his choice of problems to tackle and may leave a scientist more or less likely to hit upon a particular interpretation of what is observed. Facts, however, are supposed to be true statements about nature, statements impervious to sociological distortion. The norms of science supposedly protect scientists from subjective influences during the pursuit of the ultimate goal, objectively confirming or refuting theory. Facts may not be sufficient to determine conclusively more general truths, but empirical findings themselves have been little questioned.[22]

Under question now and here is the image of objective science, the purported objectivity of empirical fact itself—not the ambiguity inherent in the interpretation of observations, but the recorded observations themselves, the raw measurements of science. Under question is whether the heat of the scientific method has proven universally adequate to the task of sanitizing reported results, or whether in some instances observations by scientists as well as theories constructed upon observations reflect personal biases of the measurers.

The image of objective science, however necessary as guarantor of human progress, is fading. With it fades a barrier to a fuller understanding of science and civilization. In its place rise opportunities for further progress in understanding man and his works.

2

Believing Is Seeing

THE CONSCIOUS QUEST by scientists for objectivity is a distinguishing characteristic of modern science, and a characteristic that has but slowly emerged from the practice of science since the Scientific Revolution of the sixteenth and seventeenth centuries. Initially, as theories became more precise and even precisely quantitative in some instances, scientists came to know what there was to see with a greater certainty than could be supported by their new and still relatively rudimentary instruments. For some astronomers of the seventeenth century, believing became seeing.

An especially striking example of the transformation of belief into purported observation is found in Thomas Harriot's drawing of the lunar landscape. Harriot had studied at Oxford and then accompanied Sir Walter Raleigh to Virginia in 1585 as the expedition's cartographer. When Raleigh lost favor with Queen Elizabeth, Harriot found a new patron in Northumberland. It was out of the frying pan and into the fire when Northumberland participated in the Gunpowder Plot to blow up the royal family and the Houses of Parliament. King James himself set the questions for Harriot's interrogation, primarily to learn if Harriot had cast the King's horoscope. Soon Harriot was out of the Tower, and with a pension from Northumberland.[1]

With the newly discovered telescope, Harriot on 26 July 1609 observed the moon and produced a drawing (Fig. 2.1A). This drawing is the earliest extant record of an astronomical observation

Fig. 2.1 A, B, C, D. Harriot's first drawing of the moon (A), from his observation of 26 July 1609, compared with an actual photograph of the moon (B). The crescent ends in Harriot's drawing extend impossibly far down and the shading is haphazard. Harriot's second lunar drawing (D), of July 1610, is a great improvement, but it more closely duplicates Galileo's drawing (C) than the actual lunar surface. Idiosyncratic items in Galileo's drawing and in Harriot's second drawing include the large round crater close to the terminator (the dividing line between the lit and unlit hemispheres), the three little arms pointed into the dark hemisphere, and the small craters on the upper right-hand side of the moon. Harriot's ability to see depended more on his knowledge of Galileo's drawing than on his own telescope.
Source: Harriot's drawings are from manuscripts at Petworth House and are reproduced with the permission of Lord Egremont. The photograph of the moon was taken at the Lick Observatory. See Terrie Bloom, "Borrowed Perceptions: Harriot's Maps of the Moon," *Journal of the History of Astronomy* 9(1978):117–22.

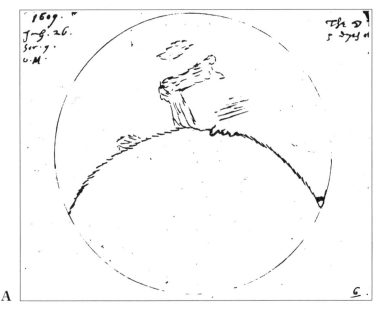

A

B

C

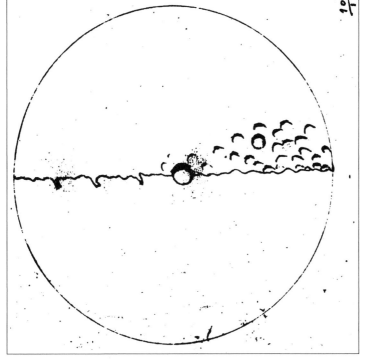

D

made with a telescope but possesses little additional merit. Made with a poor telescope magnifying only six times, the drawing is scarcely a recognizable representation of the lunar landscape. The shading is haphazard, with no suggestion of craters and mountains; and the terminator, the line between the light and dark sides of the moon, curves in an impossible manner.

Harriot's second drawing, made nearly a year later on 17 July 1610 (Fig. 2.1D), shows great improvement. Craters and mountains are now clearly delineated, and the terminator is convincingly presented. Improvement in the drawing was not, however, due to improved instrumentation, nor to improved observing skill. Improvement was an artifact of Harriot's mind, stimulated and nourished by knowledge of what Galileo had seen of the moon's surface during the intervening year (Fig. 2.1C).

It is evident from correspondence that Harriot and his friends initially were puzzled by the strange spottedness over the moon and had no idea that they were seeing shadows cast by craters and mountains. By the summer of 1610, though, they were discussing Galileo's reported observations. At first the discussion probably was based on hearsay, but Harriot could have seen a copy of Galileo's book *Sidereus Nuncius*, or the *Starry Messenger*, by early July 1610. That Harriot's July 1610 drawing of the moon contains several idiosyncratic features (Fig. 2.1C), all of which are found also in Galileo's drawings and none of which are found in modern photographs of the moon (Fig. 2.1B), argues conclusively that the origin of Harriot's drawing lies in Harriot's mind rather than in his telescope.

A comparison of Harriot's drawing, one of Galileo's drawings, and a good photograph of the moon finds that both Galileo and Harriot but not modern photography picture an enormous round crater near the center of the terminator, three jagged protrusions extending from the bright hemisphere across the terminator into the dark side of the moon, and a field of craters on the right-hand side of the moon adjoining the terminator and on the bright side. "It can even be debated," concludes Terri Bloom, who wrote an undergraduate paper on this topic, "as to whether Harriot's map bears a closer resemblance to Galileo's or to the real moon."[2] Bloom's statement is understated.

Nor may Galileo's drawings have been entirely an artifact of observation, entirely free of mental assistance. Owen Gingerich, an astronomer, historian of astronomy, and Bloom's teacher, believes that the field of craters drawn by Galileo could have been but fuzzy images seen through the telescope and were clearly seen only in the mind's eye, aided by the larger, recognizable crater on the terminator. Gingerich also notes that the size of the enormous round crater drawn by Galileo near the center of the terminator may be commensurate with the psychological impact on Galileo's thinking of Albategnius, an actual and considerably smaller crater.[3] Galilean expert Stillman Drake, on the other hand, suggests that Galileo was obliged to depict in his small engraving the crater much larger than it actually was to show the contrasting illumination of the rim at first and last quarter.[4]

Only rarely were early telescopes powerful enough for naive, unbiased observers to recognize what they were seeing. The ability to see often was dependent less on visual acuity and more on the enhancing effect of theoretical preconception.

The effect of preconception on perception is experienced by historians of science as well as by scientists in the historical interpretation of facts as well as the seeing of facts, and Harriot's sunspot observations have been reinterpreted. Bloom, fresh from the lunar study, readily ascribed Harriot's improved sunspot observations to preconception rather than to improved observing skill.[5]

Harriot first depicted sunspots as discrete dark splotches. As in the lunar case, Harriot initially had but a vague perception of what he was seeing, unaided by other astronomers' descriptions. Nor did he formulate opinions regarding either the nature or the philosophical significance of the phenomenon. Then, as also in the lunar case, there occurs a time gap of about a year from Harriot's first sunspot observations in December 1610 and January 1611 to the commencement of a systematic sunspot observation program on 1 December 1611. With the renewal of observations, Harriot more accurately depicts sunspots as nebulous blemishes.

Bloom believes that for the later reports Harriot was inspired by Fabricius's *De Maculis in Sole Observatis,* which asserted that sunspots were cloudlike and on the surface of a rotating sun. This seems more likely than attributing Harriot's improved observing skill to a long series of sunspot observations now lost; the disappearance of sunspot records but not other observing records for much of 1611 is improbable. A year-long suspension of curiosity, or at least a suspension of curious observations of sunspots, also is difficult to accept—but not impossible, inasmuch as a similar suspension seemingly occurred in the lunar case.

Prior to publication of Bloom's study, the historian of science J. D. North had suggested that the improvement in Harriot's sunspot observations might have resulted from observations during 1611 of which no record survived.[6] The extant sunspot records may be fair copies rather than rough observing notes; thus, perhaps, the gap in the observations can be explained away by the destruction of all original records and the survival of copies only.

Differences between originals and fair copies and between pencil and ink may also be relevant to an analysis of Harriot's observations. Original drawings in pencil might have shown more detail than fair copies made with pen and ink.[7] Or, if the extant drawings are originals, the medium of ink may not have allowed as much artistic expression or have admitted of as much shading as would have pencil.[8]

Neither Bloom's nor North's suggestions regarding the origin of the improvement in Harriot's sunspot observations are decisive, in contrast to the conclusive nature of Bloom's argument in the lunar case. The sunspot observations are suspect, but not decisively proven plagiarisms, at least not yet.

Early telescopic observations were qualitative, and particularly subject to subjective factors. Harriot's lunar drawing furnishes a particularly good example of the transformation of belief into see·ing. Soon telescopes would produce quantitative measurements. Here, too, believing became seeing, especially in the case of Robert Hooke and the stellar parallax.

Stellar parallax is the changing angular position of a star observed from the earth in motion around the sun. Inability to detect a measurable stellar parallax was a criticism of Copernicus's theory. Copernicus replied that stellar parallax existed but was too small to measure because the stars are many times more distant from the sun and the earth than previously imagined.[9] Among astronomers staking their claims to honor upon the first purported measurement of stellar parallax was Robert Hooke.

Hooke's career and livelihood were based and dependent upon scientific achievement. In this sense he was a professional scientist, one of the first. More common in England during the seventeenth century was the independently wealthy amateur scientist. Hooke had been left fatherless at the age of thirteen, with an inheritance of but a hundred pounds. Already safely entered in Westminister School, his abilities were such as to lead him to Oxford. There his aptitude for devising and performing scientific experiments won him a post of assistant to the affluent scientist and adventurer Robert Boyle. Upon the restoration of Charles II, some of the Oxford scientists moved to London, where they helped create the Royal Society. Hooke became its curator of experiments, obliged to furnish "three or four considerable experiments" for the entertainment of the members at their weekly meetings. There would be a small salary, when the society could afford it. Two lecturing appointments also helped sustain Hooke.[10]

"Whether the Earth move or stand still hath been a problem, that since Copernicus revived it, hath much exercised the wits of our best modern astronomers and philosophers . . .," wrote Hooke in his 1674 *An Attempt to Prove the Motion of the Earth by Observations.*[11] Well aware of the parallax problem posed by Copernican theory, Hooke also presumably appreciated the fame that would accrue to the first to prove the motion of the earth.

Encouragement from the Royal Society also spurred Hooke on in his efforts to capture the stellar parallax, a stimulus which Hooke may himself have provoked. In reports of the activities of the Royal Society there appear notes from 20 June 1666 through 22 October 1668 stating that Hooke "undertook to make observations of the parallax of the earth's orb to seconds," and "Mr. Hooke renewed his former proposal of observing the parallax of the earth's orb; which he was exhorted by the President to do with all convenient speed," and "Mr. Hooke was ordered, during this vacation, to make the experiment in the park for the mensuration of the earth; and that of observing the parallax of the earth's orb," and yet again "The president of the Royal Society mentioned, that he had understood, that Mr. Hooke had erected a tube telescope to try, whether he could observe to a second minute the passing of any fixt stars over the zenith, and thence find a parallax of the earth's orb, in order to determine the earth's motion."[12]

Evident in reports of meetings of the Royal Society is not only an increasing pressure upon Hooke to detect the stellar parallax, but also hints from Hooke that he was faithfully carrying out his assigned duty. According to the minutes for 15 July 1669, "Mr. Hooke intimated, that he was observing in Gresham-College the parallax of the earth's orb, and hoped to give a good account of it." He had, on 6 and 9 July, made his first observations of the star γ Draconis to prove parallax, though Hooke reported no numerical values. On 6 August and 21 October he made two more observations of γ Draconis, again without reporting any numbers.[13] Following the 15 July 1669 meeting, the Royal Society was not to hear again from Hooke on the matter of the stellar parallax for more than a year. Then, on 28 July 1670:

Mr. Hooke reported to the society, that he had already found so much, as to suspect some parallax of the earth's orb, and conceived, that it would be more sensible half a year later. He said, that by a perpendicular tube he observed the stars, which pass our zenith, at different times of the year, and by noting, whether the same star be at those different times of observation at the same distance from the zenith or not; concerning which he affirmed, that a certain star was then less distant from the zenith than it had been a month before.[14]

Again, no numerical values.

Hooke's intimation of success intensified the exhortation to measure the stellar parallax, and later minutes read: "He was desired to prosecute carefully this observation, so important to determine the controversy concerning the motion of the earth," and "recommended to Hooke during this recess . . . To continue to observe, whether there be a parallax in the earth's orb," and "He was likewise exhorted to prosecute the observation of the parallax of the earth's orb; concerning which he said, that he thought indeed he should find a parallax, unless it be said, that there may be a variation in the perpendicularity."[15]

Everything suggests that Hooke intended to continue his attempt to detect stellar parallax, but he was to obtain no further measurements after the four of γ Draconis in 1669 before he published in 1674 on the order of the Royal Society his tract on the attempt to prove the motion of the earth. There were many diversions, including Hooke's efforts to improve telescopes.[16] He stated that "inconvenient weather and great indisposition in my health, hindered me from proceeding any further with the observation. . . ."[17] In his tract Hooke concluded, again without reporting any numerical values, "that there is a sensible parallax of the Earths [sic] Orb to the fixt Star in the head of *Draco,* and consequently a confirmation of the *Copernican System* against the *Ptolomaick* and *Tichonick.*"[18]

Hooke's purported *experimentum crucis* attracted little attention probably because no one believed Hooke's published claim, suggests the astronomer and historian of astronomy Don Fernie.[19] Henry Oldenburg, the secretary of the Royal Society, described in his voluminous correspondence many of Hooke's accomplishments; but Oldenburg did not mention the purported parallax measurement. Nor did the English astronomer John Flamsteed bother to mention Hooke when claiming credit a few years later for his own measurement of stellar parallax.[20]

Hooke may not have had much more confidence in his own claim than his contemporaries had; he seems not to have continued his parallax work beyond 1669. Hooke did, however, repeat his claim once. In 1691 the French astronomer Jean Cassini wrote that some English astronomers had found evidence that the annual orbit of the earth was not insensible in comparison to the distance of the fixed stars, but the cause of the observations was not yet evident to him.[21] Hooke believed that the French were angry that they had not first thought of attempting to measure stellar parallax, and in 1692 he repeated his unsubstantiated claim to have discovered the parallax of the earth's orb.[22]

In the 1720s the English astronomer James Bradley began observing γ Draconis, intending to verify Hooke's parallax measurement. Instead, Bradley found and correctly interpreted evidence of stellar aberration, the apparent displacement of a star in the direction in which the earth is moving.[23] (The effect is often illustrated with raindrops and a moving person holding an umbrella.)

The stellar aberration is larger than Hooke's parallax value, and it would not be unreasonable in the absence of other information to attribute Hooke's claimed stellar parallax to the presence of stellar aberration. However, as Bradley noted, the direction in which Hooke's purported parallax supposedly displaced γ Draconis is opposite to the effect produced by stellar aberration. There is no excuse other than ignorance for modern attributions of Hooke's alleged parallax to the effect of stellar aberration. Nor could Bradley find any other explanation for Hooke's error.

As to the observations of Dr. *Hook* [sic], I must own to you, that before Mr. *Molyneux's* Instrument was erected, I had no small opinion of their Correctness; the Length of his Telescope and the Care he pretends to have taken in making them exact, having been strong Inducements with me to think them so. And since I have been convinced both from Mr. *Molyneux's* Observations and my own, that the Doctor's [Hooke's] are really very far from being either exact or agreeable to the *Phaenomena;* I am greatly at a Loss how to account for it. I cannot well conceive that an Instrument of the Length of 36 Feet, constructed in the Manner he describes his, could have been liable to an Error of near 30″ (which was doubtless the case) if rectified with so much Care as he represents.[24]

Stellar aberration does not explain Hooke's error. Nor were there complications in mathematical reductions of the data. Nor was refraction (or bending) of light in the earth's atmosphere a problem because Hooke had chosen γ Draconis, a star near the zenith (directly overhead), to avoid the problem of refraction, which is great for objects near the horizon and of little matter for objects at the zenith. Hooke might have been extraordinarily careless, but for some of his measurements using telescopic sights he did claim an accuracy of nearly a second of arc. A contemporary of Hooke's, John Hevelius, thought that he could measure accurately to a fifth or even a tenth of a minute without telescopic sights.[25] It would require a random error of nearly 30 seconds of arc, as Bradley noted incredulously, to explain away Hooke's purported parallax as a random error. It is difficult to believe that Hooke was so careless. Easier to imagine is Hooke, unconsciously influenced by preconception and spurred on by pressure from the Royal Society, straining in the summer and fall of 1669 to detect a quantity slightly beyond his observational ability.

Hooke's parallax measurement was not accepted by contemporaries, but Flamsteed's was. As the first Astronomer Royal at the Royal Observatory at Greenwich, Flamsteed's major task in 1675 was to measure with the greatest accuracy possible the positions of stars. In a letter to the bishop of Salisbury in 1680, Flamsteed described an effect he had found and asserted that it could only be the sought-for parallax. He did not, however, pursue this claim. Probably it was an effort to impress the bishop with an eye to future patronage, so the historian of science Mari Williams believes.[26] Flamsteed's second claim was more convincing. In the 1690s he took up again the parallax problem, this time with a new mural arc 7 feet in radius, which reduced by half the margin of error. From observations of the pole star made over a period of five or six years, Flamsteed detected a systematic difference between the position of the star in summer and in winter. He attributed the difference to parallax, informed friends as early as 1695, and published his result in 1699. It was impressive to many, especially because of the care that Flamsteed had taken in correcting his observations for atmospheric refraction and for the subsidence of the wall to which the mural arc was fixed.

Flamsteed's result was not impressive, however, to the French astronomer Jacques Cassini, son of the Jean Cassini who had earlier criticized Hooke's work and the second of four generations of Cassinis to direct the Paris Observatory. Cassini *fils* had spotted a fatal flaw in Flamsteed's report. Flamsteed's understanding of stellar parallax was faulty, and his imperfect theory incorrectly predicted that the pole star would be closer to the pole in winter than in summer—precisely what he claimed to have observed. Made aware of his error in theory and prediction, Flamsteed proceeded to blame the corresponding error in observation on some unknown instrumental error. He was never able, though, to identify the source of the error. Williams labels Flamsteed's observation "a clear example of someone finding something he was determined to find."[27]

Flamsteed, Hooke, Harriot, and perhaps Galileo had seen more than was to be seen through their new instruments unaided by imagination. Early professional scientists in a field just emerging, they were prey to the pressures of performance and patronage, with neither the protection afforded by an awareness of the effect of preconception nor the protection afforded by procedures yet to be developed to check the effect of preconception upon observation. Such protection might not prove 100 percent effective; it would not spare all twentieth-century scientists the embarrassments of their seventeenth-century predecessors. Still, in the quest for objectivity if not necessarily in the attainment of objectivity, seventeenth-century science is a marker against which significant subsequent progress can be measured.

3

Planetary Fantasies: Uranus

EARLY EXCESSES of astronomical enthusiasm in the seventeenth century associated with newly visible objects in new instruments and newly expected phenomena predicted from new theories did not fade away as astronomy matured. Expectations intensified with increasing knowledge; the transformation of belief into vision continued. Three particularly striking planetary fantasies occurred at the end of the eighteenth century, in the middle of the nineteenth century, and at the beginning of the twentieth century.

That is not to say, though, that the quest for objectivity was stillborn or never even conceived. The planetary fantasies were recognized as such, even by some of the participants, and stimulated much conscious awareness and examination of the dangers posed by preconception to objectivity. Indeed, William Herschel's work late in the eighteenth century constitutes a major if little appreciated early step in the quest for objectivity.

One of the most exciting discoveries in astronomy was Herschel's sighting of the planet Uranus on the night of 13 March 1781. For millenia, Mercury, Venus, Mars, Jupiter, and Saturn had been known to man; now Uranus joined them and the earth as a seventh planet. Or not quite "now," because Herschel at first harbored not the faintest suspicion that he had discovered a new planet. It wasn't until the Astronomer Royal, Nevil Maskelyne, remarked upon the object's planetlike appearance and French astronomers calculated a planetlike orbit that the true nature of

Uranus became known. In the meantime, Herschel was generating quantitative measurements of the apparent diameter of Uranus and of its position in the skies, data now known to have been a product of his belief rather than his vision.

Herschel had left the Hanoverian Guards' regimental band in 1757, following a disastrous encounter with the French army, and moved to England with his older brother Jacob. Early desertion and later pardon from the hands of King George III himself in honor of Herschel's discovery of Uranus is a beautiful story, spoiled only by the ugly fact of an official discharge dated 29 March 1762, well after the fact of Herschel's questionable departure but undisputedly many years prior to the discovery of Uranus.[1]

Newly arrived in London with not half a guinea in the world, Jacob looked for engagements to play the violin and did some teaching while William added to their meager income by music copying. Jacob returned to Hanover in September 1759 and William, realizing that London was relatively crowded with musicians, set out in 1760 to try for better success in the country. There followed a brief stint as bandmaster of a militia regiment quartered at Richmond in Yorkshire and a succession of free-lance engagements, composing, directing, and playing, at Sunderland, Newcastle, Edinburgh, and Pontefract. It was not an entirely satisfactory life, as Herschel indicated in a letter to his brother in 1761:

> You don't perhaps know that, I have already some time been thinking of leaving off professing Musik and the first opportunity that offers I shall really do so. It is very well, in your way, when one has a fixed Salary, but to take so much for a Concert, so much for teaching, and so much for a Benefit is what I do not like at all, and rather than go on in that way I would take any opportunity of leaving off Musik; not that I intend to forget it, for it should always be my chief study tho' I had another employment. But Musik ought not to be treated on that mercenary footing.[2]

Life took a turn for the better the next year when Herschel won appointment as director of public concerts in Leeds. He remained there for four years, was briefly organist at Halifax in 1766, and took up the appointment of organist at the Octagon Chapel in Bath on 9 December 1766.

Music directed Herschel's interest to the subject of harmony, which in turn led him to mathematics and to astronomy. Robert

Smith's text *Harmonies, or the Philosophy of Musical Sounds* led Herschel on to Smith's other book, *System of Optics*. By 1766 astronomy was for Herschel a marginal hobby. By 1779 it was a serious hobby; Herschel cut back the number of musical pupils so that he could devote more time to constructing telescopes and observing the heavens.

Observing with a telescope of his own construction on Tuesday, 13 March 1781, Herschel noticed an object visibly larger in appearance than nearby stars. And its image increased in size with higher magnification. From experience Herschel knew that the diameters of stars are not proportionally magnified with higher powers, as are the diameters of planets, comets, and other objects with extended images. He recorded in his journal the discovery of "either a nebulous star or perhaps a comet."[3]

If a nebulous star, it would remain fixed in its position; if a comet, it would be seen to move across the sky. The choice was delayed for three nights, presumably by cloudy skies. On Saturday, Herschel "looked for the Comet or Nebulous Star and found that it is a Comet, for it has changed its place."[4]

Herschel began regular observations of what he presumed to be a comet, and on 25 March noted that "the apparent motion of the Comet is accelerating, and its apparent diameter seems to be increasing." Three days later he was more sure: "the diameter is certainly increased, from which we may conclude that the Comet approaches us." A series of measurements from 17 March through 18 April of the size of the apparent comet supposedly coming nearer the earth purportedly documented a gradual increase in the apparent diameter of the object:

Date	Diameter
March 17	2″3‴
19	2 59
21	3 58
28	4 7
—	3 58
29	4 7
—	4 25

April	2	4″25‴
	6	4 53
	15	5 11
	—	5 20
	18	5 2

Note: ″ = second of arc; ‴ = 60th of a second.[5]

Herschel observed a gradual increase in apparent size. What he should have observed was a gradual *decrease* of Uranus's apparent diameter, because in March 1781 Uranus was moving not toward but away from the earth, as R. H. Austin, a historian of science, has noted. A similar relationship for the positions of earth, sun, and Uranus occurred in March 1950. Then the change in apparent diameter was:

Date	Diameter
March 17	3″38‴
22	37
28	36
April 3	35
April 10	34
April 17	32

Note: ″ = second of arc; ‴ = 60th of a second.[6]

Initially Herschel could not have known, independently of his own measurements, what the diameter should have been. After the calculation of Uranus's orbit, however, the fault in Herschel's measures was evident. Thomas Hornsby, professor of astronomy at Oxford, where he would promote construction of the Radcliffe Observatory, wrote to Herschel on 26 February 1782 concerning several problems in the observations of Uranus, including the fact that when Herschel "first discovered the Comet . . . its Distance from the Earth was increasing. . . ."[7]

Herschel had also anticipated a cometlike path for the supposed comet. When inconsistencies between expectation and observation arose, he attributed them to the motion of the earth, which changes the observer's vantage point and hence angles measured. Herschel added a caveat to this section of his work: a very small error in the measures would produce a very large error in the distance, and the parallax observations hence were extremely precarious. Later Herschel omitted from his published report the entire discussion of the parallax measurements.[8]

Astronomers were justly skeptical of Herschel's measurements. In addition to Hornsby, Maskelyne especially doubted that Herschel could measure astronomical positions with the accuracy he claimed. Yet the doubts were voiced only privately, so softly shouted that their first public hearing came only on the two-hundredth anniversary of the event, in a study by Simon Schaffer, a historian of science. More attention was devoted by eighteenth-century astronomers to establishing a new observational astronomy in England based on Herschel's work than was devoted to a critique, constructive or destructive, of Herschel's achievement. Questions about the reliability of Herschel's observations would not have enhanced the attempt to legitimate in public and to obtain the king's patronage for the way Herschel used his instruments. The effort was successful; George III came to prefer expending funds on the study of astronomy rather than on the killing of men.[9]

One response, once Herschel's measurements proved embarrassing, was an attempt to explain them away as instrumental errors rather than the result of preconception. In doing so, Herschel also made some progress toward the identification of effects of preconception. He preserved the image of a successful new astronomy and also added to its substance. The latter endeavor is, ultimately, as necessary as the former for the advance of science.

Perhaps it was the privately expressed concerns of Hornsby and others that prompted Herschel to read before the Royal Society on 7 November 1782 a paper on Uranus. In the paper Herschel discussed the inconveniences and deceptions to which almost every sort of micrometer is liable, although he did not address explicitly the issue of the erroneous measurement of an increase in apparent diameter. Recognizing more or less explicitly the effect of preconception upon observation, Herschel now set aside his first three measures of the apparent diameter of Uranus on the supposition "that every minute object, which is much smaller than what we are frequently used to see, will at first sight appear less than it really is."[10] Herschel had already noted this phenomenon in a footnote to his original paper on Uranus.[11] Now he elaborated upon it and used it as justification to throw out the first three measurements. A mean of the remaining measures gave a diameter for Uranus of 4″ 36½‴, with all but two of the measures within half a second of the mean. Herschel thus emphasized, perhaps somewhat disingenuously, the consistency of his measures about the mean rather than the consistent increase in size he had reported earlier but now avoided mentioning. Herschel also searched for errors possibly introduced in the measures by his micrometers.[12]

Herschel's sophistry perhaps can be applauded, or at least understood, in the context of the effort to establish a new observational astronomy in England.[13] Early errors publicly perceived might have crippled or even killed the enterprise. However dubious the defense, the end perhaps helps justify the means.

The defense was dubious. Herschel did not satisfactorily explain away his initial and erroneous measurement of a steadily increasing apparent diameter. As Austin has argued, any explanation of Herschel's error relying on instrumental factors or even sheer carelessness must be incomplete, because random errors would not produce a gradual and consistent increase in the measured diameter.[14] Nor could random errors have produced a consistently too-small diurnal parallax.

The influence of expectation upon observation must be part of any explanation of Herschel's erroneous observations. Austin notes that as a general rule, "the scientist's interest in his work will lead him to formulate hypotheses and the hypotheses will lead him to expectation about future observation." Under these conditions, Austin asks, "is disinterested investigation possible?" Scientists are on guard, "but in cases where discriminations are exceptionally difficult to make, then here, perhaps, in spite of pre-knowledge of the danger, and in spite of all precautions, an expectation may influence a judgment."[15]

Clearly Herschel had expected to measure an increasing apparent diameter, though when precisely he formed the expectation is not certain. He may have held that expectation from the beginning, since comets generally are discovered approaching the earth. Austin suggests, however, that Herschel may have decided the object was approaching the earth only after his first two measurements.[16] The first measurement would have been too small, since minute objects smaller than expected initially appear even smaller, while the second measurement would have been more accurate and thus have created the impression of an increasing diameter. As Austin notes, Herschel by 29 March was expecting his measures to be increasing, because he labeled the 4″ 7‴ measure "rather too small a measure" while he noted that the 4″ 25‴ measure "seems right."[17]

Preconception may also have played a role in Herschel's measure of the oblateness of Uranus, the flattening at its poles due to the spin about its axis. According to Austin, it did, with diametrical result.[18] That conclusion, however, is doubtful.

On the night of 13 October 1782 Herschel viewed with several powers of magnification the Georgium Sidus—as he had named the new-found planet, conveniently conveying the time and country of its discovery, expressing his gratitude as a subject of the best of kings, the liberal protector of every art and science, and no doubt also improving his position as a dependent of that king's unlimited bounty.[19] About the planet and his observations of it, Herschel wrote: "I viewed the Georgium Sidus with several powers. With 227 it was beautiful. Still better with 278. With 460, after looking some time, very distinct. I perceived no flattening of the polar regions, to denote a diurnal motion; though, I believe, if it had had as much as Jupiter, I should have seen it. With 625 pretty well defined."[20] A month later Herschel again perceived no flattening of the polar regions.[21]

Actually, Uranus is more oblate, more flattened at the poles, than is Jupiter, as Austin has pointed out in his analysis of Herschel's observations. Austin speculates that Herschel failed to detect the flattening because preconception directed his attention away from the right place and to the wrong place. For most planets the axis of rotation is approximately perpendicular to the plane of the planet's orbit of revolution around the sun. Thus Herschel likely would have expected to find the poles and the flattening at the poles at the parts of the visible disc of Uranus most nearly above and below the plane of its orbit. Uranus's axis of rotation, however, lies nearly in the plane of its orbit. Here the evidence of flattening lay, presumably overlooked by Herschel because he expected to find it elsewhere.[22]

More evidence is needed, however, to justify indictment of preconception in the matter of the overlooked oblateness. The orientation of Uranus's axis, and hence its poles, varies. Sometimes from earth a pole is seen straight on; at other times the poles are seen left to right nearly in the plane of the planet's orbit. The period of rotation of the axis is 42 years. Uranus presented maximum side view in 1756 and 1798, and was pole-on in 1777.[23] In side view, flattening is most visible; pole-on, however, flattening is invisible to an earth-bound observer. Herschel, observing on 13 October 1782 near the pole-on alignment, could not have detected oblateness. Austin's charge against Herschel in this instance is not proven.

Questions about Herschel and the flattening of Uranus are not, however, unreasonable. As early as 8 April 1783, less than five months after his perception of no flattening and while Uranus was still aligned so that flattening could not be detected from the earth, Herschel did surmise a polar flattening.[24]

Yet more suspicious in terms of the possible effect of preconception is Herschel's later observation on 26 February 1794 that "the planet seems to be a little lengthened out, in the direction of the longer axis of the satellites' orbits."[25] Some years earlier Herschel had discovered two satellites circling Uranus and had determined their orbits to be nearly perpendicular to their parent planet's orbit around the sun, rather than parallel, as are the orbits of the satellites of Jupiter and Saturn.[26] Jupiter's and Saturn's satellites are in nearly equatorial orbits; hence Herschel might have supposed that Uranus's satellites also circled its equator. Equatorial lengthening, the converse of polar flattening, might therefore be expected to coincide with the orbital plane of Uranus's satellites. This new preconception could have rotated Herschel's attention by approximately 90 degrees from where it had been fixed by prior preconception during his search for polar flattening and equatorial lengthening. Herschel now found the lengthening. The satellites' orbits and Uranus's equator, the plane of the lengthening, were in conformity; the analogy with Saturn and Jupiter held. Later, in 1797, Herschel would write:

> The flattening of the poles of the planet seems to be sufficiently ascertained by many observations. The 7-feet, the 10-feet, and the 20-feet instruments, confirm it; and the direction pointed out Feb. 26, 1794, seems to be conformable to the analogies that may be drawn from the situation of the equator of Saturn, and of Jupiter.
>
> This being admitted, we may without hesitation conclude, that the Georgian planet also has a rotation upon its axis, of a considerable degree of velocity.[27]

The emphasis in 1797 on analogy with Jupiter and Saturn is explicit and strong. It suggests that analogy rather than observation may have been Herschel's primary guide. The analogy was not made explicitly in the 1794 report but it is there implicitly, revealed by the emphasis upon equatorial lengthening rather than polar flattening. Polar alignment is the normal issue; orbiting satellites would divert attention to the equator.

Preconception was present. Yet the purported observation occurred within four years of the time of maximum visible oblateness in 1798. The physical phenomenon may have been barely detectable. And Herschel's reported inclination of about 77 degrees for the equatorial lengthening is reasonably close to the modern value of 82 degrees (technically, 98 degrees). Herschel's early report has not been impeached by knowledge subsequently gained.

While the purported increase in the apparent diameter of Uranus is a clear case of preconception predominating, the measures of Uranus's oblateness are more difficult to classify. Probably Herschel initially did expect to find flattening at the top and bottom of Uranus's visible disc, not at the sides, and looked for flattening at the top and bottom. Speculation on whether preconception caused him in 1782 thus to overlook flattening at the sides is futile, however, because the flattening wasn't then visible from the earth. By 1794 another preconception directed Herschel's attention to lengthening rather than flattening, and to an inclination of approximately 90 degrees. This time the phenomenon was approximately there to observe. Thus no decisive attribution of Herschel's observation to preconception rather than to observation is possible, however much the switch in language from flattening to lengthening and the explicit mention of Saturn and Jupiter suggest the presence, and possibly influence, of expectation.

Also not facilely categorized are Herschel's tentative, on-again off-again suspicions of the existence of a ring or rings encircling Uranus. His ring observations do not furnish an undisputable, unambiguous example of the effect of preconception and the failure of objectivity. They do, though, well illustrate difficulties in observing and the consequent opportunity for preconception to play a role.

The ring observations are also important because in the course of the observations Herschel instituted increasingly sophisticated if simple precautions to eliminate possible instrumental errors. These are the forerunner of similar precautions exercised without complete effect in more recent cases involving preconception and failed objectivity. Basically, comparison of two objects eliminates instrumental error as the source of any feature not common to both — at least in principle if not always in practice. If done without knowledge of which object is which, the procedure should eliminate the possibility of personal bias influencing the observation. This is why medical studies are done "double blind"; neither the subjects nor the doctors reporting the subjects' reactions know which subjects have received placebos and which have received real drugs. A double-blind procedure was not feasible for Herschel's observations, nor is there any indication that he even imagined such a procedure. Herschel did, though, make comparisons as a check on his observations of possible rings around Uranus.

No doubt Saturn's ring inspired the search for a similar Uranian appendage. Instrumental improvements also furnished encouragement; only after Herschel's 1787 discovery of the two brightest satellites of Uranus with his new 20-foot reflecting telescope are a possible ring or rings mentioned in his observing notes. On 4 February 1787 Herschel saw Uranus "well defined; no appearance of any ring." A month later, however, he reported: "I begin to entertain again a suspicion that the planet is not round. When I see it most distinctly, it appears to have double, opposite points. . . . Perhaps a double ring; that is, two rings, at rectangles to each other." And the next night: "The Georgian Sidus not being round, the telescope was turned to Jupiter. I viewed that planet

32

with 157, 300, and 480 [magnification], which showed it perfectly well defined. Returning to the Georgian planet, it was again seen affected with projecting points. Two opposite ones, that were large and blunt, from preceding to following [west to east]; and two others, that were small and less blunt, from north to south."²⁸ The comparison of Jupiter and Uranus is a simple matter, when both are simultaneously visible, but important for showing that the projecting points observed were not an artifact of the telescope. Such a procedure was by no means standard practice in Herschel's time, though the thought is found even earlier. Galileo, answering criticism that his telescope put into the sky things not really there, particularly satellites circling Jupiter, offered 10,000 scudi to anyone who made a telescope that would create satellites around one planet and not around others.²⁹

Herschel soon found a great and a small ring, only to conclude the very next night, without explanation, that "R and r are probably deceptions." Later in the year Herschel noted that "the suspicion of a ring returns often when I adjust the focus by one of the satellites, but yet I think it has no foundation." The suspicion grew, and on 22 February 1789 Herschel stated flatly "a ring was suspected." More observations strengthened the suspicion. On 16 March 1789 Herschel wrote:

7ʰ 37′ I have turned my speculum [telescope mirror] 90° round. A certain appearance, owing to a defect which it has contracted by exposure to the air since it was made, is gone with it; . . . but the suspected ring remains in the place where I saw it last.
7ʰ 50′ Power 471 shews the same appearance rather stronger. Power 589 still shews the same. *Memorandum.* The ring is short, not like that of Saturn.³⁰

Earlier, Herschel had observed both Uranus and Jupiter to assure that the suspicion of a ring around one was not also produced by the telescope around the other planet. Now, he more simply rotated the telescope mirror, to see if the suspicion of a ring was rotated too.

It was not, and Herschel grew more confident. On 20 March he noted "when the satellites are best in focus, the suspicion of a ring is strongest.' And near year's end he found that "the planet is not round, and I have not much doubt that it has a ring." Two years later, on 26 February 1792, Herschel again thought he had a ring. Observations later that same night, however, raised the possibility of an instrumental defect:

6^h 34′ My telescope is extremely distinct; and, when I adjust it upon a very minute double star, which is not far from the planet, I see a very faint ray, like a ring crossing the planet, over the centre. This appearance is of an equal length on both sides, so that I strongly suspect it to be a ring. There is, however, a possibility of its being an imperfection in the speculum [mirror], owing to some slight scratch: I shall take its position, and afterwards turn the speculum on its axis.
8^h 39′ Position of the supposed ring 55°.6 from N.P. to S.F. [north preceding, or north-west, to south following, or south-east].
9^h 56′ I have turned the speculum one quadrant round; but the appearance of the very faint ray continues where it was before, so that the defect is not in the speculum, nor is it in the eye glass. But still it is now also pretty evident that it arises from some external cause; for it is now in the same situation, with regard to the tube, in which it was 3½ hours ago: whereas, the parallel is differently situated, and the ring, of course, ought to be so too.[31]

First, Herschel had compared two planets. Next, he more simply rotated the telescope. Finally, he sat back and let the earth do the rotating. Throughout, Herschel was conscious of problems inherent in observation, and he was seeking solutions.

A week later, Herschel "viewed the Georgian planet with a newly polished speculum, of an excellent figure. It shewed the planet very well defined, and without any suspicion of a ring. I viewed it successively with 240, 300, 480, 1200, and 2400; all of which powers my speculum bore with great distinctness. I am pretty well convinced that the disk is flattened."[32]

By December 1797 Herschel had been searching with the 20-foot telescope for a ring around Uranus for nearly fifteen years. Even had a ring been nearly edge on and thus nearly invisible at the beginning of his search, alignment changes should by now have made any ring visible. With great confidence in the observation of 5 March 1792, reaffirmed by later views of the planet, Herschel now ventured "to affirm, that it has no ring in the least resembling that, or rather those, of Saturn." Yet he hesitated to rule out a ring completely: "the observations which tend to ascertain the existence of rings not appearing to be satisfactorily supported, it will be proper that surmises of them should either be given up, as ill founded, or at least reserved till superior instruments can be provided, to throw light on the subject."[33]

Regarding the many suspicions of a ring, Herschel noted that the arrangement of his eyeglass close to the mouth of the telescope tube sometimes occasioned a visible defect. But he chose not to enter "further into the discussion of a subject that must be attended with uncertainty."[34] Less circumscribed, the historian of astronomy Richard Baum has chosen to analyze Herschel's ring observations. He finds it probable that astigmatism, a characteristic defect of the particular manner in which Herschel constructed his telescopes (with the main mirror tilted slightly to the optical axis, reflecting back to the eyepiece at the upper edge of the tube; that is, the front view) produced a slightly linear image and hence the impression of a ring around Uranus.[35] It was also the front-view arrangement that allowed in additional light, making it possible for Herschel to see the satellites of Uranus.

Herschel would not have been surprised to have found a ring around Uranus. He looked long and hard for one. Perhaps at times his preconception magnified an optical defect into more of a suggestion of a ring than an unbiased observer would have seen. At other times, as with Jupiter, which Herschel viewed without expectation of finding a ring and without the magnifying power of preconception, optical defect acting alone was insufficient to produce the suggestion of a ring. Despite occasional lapses, Herschel generally maintained a critical attitude. Ultimately, he found the case for a ring around Uranus not proven.

Even more than the cases of the changing apparent diameter and the oblateness of Uranus, the case of a possible ring illustrates how precariously balanced Herschel was on the edge of objectivity. Pressures associated with the procurement of patronage were not a stabilizing factor in maintaining the precarious balance; Herschel, to his great credit, generally resisted this distraction. Also to Herschel's credit are the several and increasingly sophisticated comparisons used to determine whether a particular phenomenon was an artifact of the instrument or a real object far distant in space. If Herschel did not achieve objectivity, he went far in the quest.

4

Planetary Fantasies: Neptune

NEPTUNE, the next planet discovered after Uranus, also illustrates the conversion of expectation into purported fact. Saturn's ring once again suggested the possibility of more rings, this time around the newly discovered planet Neptune. Furthermore, special circumstances practically demanded that those who had failed to win for England the honor of the discovery of Neptune itself should not fail a second time. If there was a ring, they must find it. Considerations of prestige as well as potential patronage pressured observers to see what they believed was there to see.

Study of Uranus had led to the discovery of Neptune. Continued observations of Uranus after its discovery by Herschel in 1781 enabled astronomers to calculate an orbit by 1790. The orbit satisfied observations made between 1781 and 1790 and also a few earlier observations that in retrospect were realized to have been of Uranus.

All seemed well—but not for long. The chaos of the Napoleonic wars was scarcely conducive to the support of scientific observations, and astronomers retreated to their libraries. There, among records of old observations, they found yet more pre-1781 sightings of Uranus. By 1820 it was obvious that the orbit for Uranus determined in 1790 did not fit all the observations now known, even after making corrections for the gravitational pull on Uranus of Jupiter and Saturn.

One solution to the apparent discrepancy between prediction from the theory of gravity and actual astronomical observation was to hypothesize the existence of an eighth planet beyond Uranus whose gravitational pull on Uranus would produce the observed orbit. Many astronomers made such a suggestion; two mathematicians calculated where the new planet should be.

Urbain Jean Joseph Le Verrier, a young French mathematician, took up the problem of Uranus's motion in 1845 after working on planetary orbits for some eight years. By November he had finished his memoir on the theory of Uranus. Early in 1846 Le Verrier was corresponding with George Airy, the astronomer royal and director of the Greenwich Observatory, about his prediction. And on 18 September 1846 Le Verrier wrote to an astronomer at the Berlin Observatory, asking him, too, to look for the predicted planet.

Meanwhile in England, independently of Le Verrier's work, John Couch Adams, a young Fellow of Cambridge University and a brilliant mathematician, had been working on the problem of Uranus's orbit since early in 1843. He completed a preliminary solution in that year and was soon at work on a better one. In February 1844 James Challis, the director of the Cambridge Observatory, wrote to Airy to tell him of Adams's work on Uranus and to ask for Airy's observations. Adams completed his new solution in September 1845 and gave it to Challis and Airy. Because of various mishaps, though, Adams did not see Airy in person. Airy was skeptical of theoretical investigations and of young men; it was not until Airy saw Le Verrier's paper that he began to take much interest in the proposed new planet. Then he moved cautiously and slowly. Too slowly.

On 9 July 1846 Airy wrote to Challis asking him to look for the planet, or if he could not, to set someone else at the observatory to the task. Challis began a systematic search for Neptune, a search that would have benefited from a desultory element. Challis unknowingly observed the planet in August and again in September before losing for England the honor of the discovery.

Indeed, on 29 September, following Le Verrier's suggestion in a letter received that same morning, Challis observed a star for which he directed his assistant to write in the observing log "seems to have a disc."[1] But Challis was not yet ready to analyze his accumulating data nor to claim the honor of discovery. And it was already too late. Johann Galle at Berlin, acting expeditiously on a letter from Le Verrier that had reached Germany on 23 September, discovered Neptune the next night. News of the discovery reached London a week after the fact, and after Challis's observation of a star seeming to have a disc. An announcement appeared in the *Times* on 1 October.[2]

Airy was savagely abused, both by the English, because one of their countrymen had not made the discovery, and by the French, who saw Airy's belated notice of Adams's work as an attempt to lessen Le Verrier's achievement, a challenge to Le Verrier's priority. Challis, too, came under heavy fire in England. He and Airy were forced to make public accountings of their roles in the fiasco.[3] Excitement over the new planet would encourage studies of its features; hurt pride was further goad to the English.

The new planet came quickly under telescopic scrutiny. One of numerous astronomers examining its image was William Lassell. A provincial brewer by trade, Lassell had recently installed at Starfield, his private observatory near Liverpool, a new 24-inch reflecting telescope, the most powerful telescope then in England. The Earl of Rosse had a 36-inch reflector in Ireland, but it was not an effective instrument. The mounting was poor and the lens both tarnished quickly and required repolishing after a few cleanings. Also, Rosse was diverted from astronomy by the Irish potato famine from 1845 to 1848, and then diverted again by his duties as president of the Royal Society from 1848 to 1853.[4] Astronomers turned to Lassell for definitive observations of Neptune.

There is a story that Lassell had known of Adams's prediction of an eighth planet as early as October 1845, nearly a year before its discovery, but had been kept from searching for the predicted planet by various circumstances. The story, however, is suspect.[5] Lassell probably did not know of Neptune until he saw in the *Times* of 1 October 1846 the announcement of the planet's discovery.

That same day Lassell received from John Herschel, the son of

William Herschel and an accomplished scientist in his own right, a letter urging Lassell to look for "satellites [of Neptune] with all possible expedition!!"[6] Seemingly Lassell expanded Herschel's suggestion to look for satellites into a suggestion to look for satellites *and* for a ring, because he soon replied to Herschel, "I am obliged by your note directing my attention to the possible ring and satellites of LeVerrier's planet."[7] It is not clear whether the expansion of Herschel's suggestion in Lassell's mind took place before or after Lassell's observation, whether Herschel's letter predisposed Lassell to expect to find a ring. The example of Saturn could also have so predisposed Lassell, as historians Robert Smith and Richard Baum point out in their study of Lassell and the ring of Neptune.[8]

On 2 October Lassell easily found Neptune, resolving its disc without difficulty. In his diary he noted: "I observed the planet last night the 2nd and suspected a ring . . . but could not verify it. I showed the planet to all my family and certainly tonight have the impression of a ring."[9] Lassell's next observation of Neptune occurred on 10 October. Bad weather, presumably, had interrupted his search for a satellite and for a ring. Now he saw "a satellite or a suspicious looking star . . . and a ring. I see the [satellite] and suspect the ring with various powers [magnification] 316 to 367. . . ."[10] And in the *Times* of 14 October 1846 Lassell reported his discovery of Triton, the faint inner satellite of Neptune: "Of the existence of the star, having every aspect of a satellite, there is not the shadow of a doubt."[11] Lassell was less certain, but nevertheless still very confident, about the existence of a ring: "I am not able absolutely to declare it, but I received so many impressions of it, always in the same form and direction, and with all the different magnifying powers, that I feel a very strong persuasion that nothing but a finer state of atmosphere is necessary to enable me to verify the discovery."[12]

More observations, through the middle of December until terminated by bad weather and the increasing distance of Neptune from earth, strengthened Lassell's impression of a ring. Observations by several other persons viewing the planet through Lassell's telescope also strengthened his belief.[13] Lassell was not unaware of the possibility of deception, and he took steps to reduce the possibility:

To guard against deception in these observations I called in the aid of several friends totally unacquainted with the impressions I had received, and I found that the diagrams they severally and independently drew agreed as nearly as the representations of the same appearance by several hands were likely to do. I further took the precaution (to prevent anything in the telescope from deceiving me) to put other specula [mirrors], both large and small, into the tube the impression of the ring still remaining. Finally I carefully scrutinized the planet Uranus without being able to persuade myself that any such appendage belonged to him.[14]

According to Lassell, his friends not only had seen a ring, but they had seen it in the same direction as he had.[15]

Reports soon followed from a few other astronomers who found themselves capable of duplicating Lassell's visual feat with the supposed ring of Neptune. As Baum has noted, some who saw no ring when they first viewed Neptune were more perceptive after Lassell's report appeared.[16]

John Hind was one. He had been an assistant at the Royal Observatory before becoming the director of George Bishop's private observatory in Regent's Park, London. On 30 September 1846 Hind wrote to Challis at Cambridge, reporting that he had "observed the planet this evening; it is bright and the disc is distinctly perceptible."[17] Hind mentioned not the least suspicion of a ring. But in December, after publication of Lassell's purported discovery of a ring, Hind came to think that the existence of a ring, while not yet decided, was most probable because the planet now seemed to present an oblong appearance in Mr. Bishop's 7-inch refracting telescope.[18]

Challis was another. Using the larger Northumberland 11³/₄-inch refracting telescope, he also came to suspect the presence of a ring—and also only after Lassell's announcement of a ring. Earlier Challis had unknowingly observed the planet on more than one occasion, sensing neither the presence of a disc nor the slightest suspicion of a ring.[19] These early observations could not have detected a ring, even were there a ring to detect; high magnification would be necessary. Regular observations of the new planet were commenced at Cambridge on 3 October 1846, regular at least insofar as the notorious English weather permitted. Neptune was hidden from Cambridge observers by clouds in late December and early January, just when Lassell's reports of a ring were appearing.

Not until 12 January 1847 did Challis obtain his first "distinct impression" of a ring surrounding Neptune.[20] He reported that he saw the ring again on 14 January and asserted that his assistant also had seen the ring with the same alignment about the planet. As an indication of the observing conditions, Challis remarked that under similar conditions he had been able to see the second division of Saturn's ring.[21]

Challis wrote to Lassell and learned that bad weather had prevented Lassell from observing Neptune for some time. Lassell had heard "somewhat indefinitely" that Francesco de Vico, the director of the Collegio Romano Observatory, had observed the planet "to have constantly two appendages which may be either a ring or a cluster of satellites."[22] De Vico's observations, whatever they were, now were unimportant thought Lassell when compared with Challis's observations. Challis's observations seemed to put "beyond reasonable doubt" the reality of the ring, "especially as even your measured angle of position agrees with my estimation within 4 degrees," so Lassell wrote to Challis.[23]

But were Lassell and Challis really in agreement regarding the orientation of the supposed ring? It is customary in drawing a planet to draw it with south up and north down, east to the right and west to the left, because the usual refracting telescope inverts the image. Stars and planets cross the meridian from east to west, *following* from the east or right and *preceding* toward the west or left. Challis drew his ring from upper left to lower right, south-preceding to north-following.[24] In his initial reports Lassell had not specified to which side of the north-south axis of the planet his ring was tilted.[25] Nor had Hind.[26] In his 19 January 1847 letter to Challis, in reply to Challis's letter of 16 January announcing his observation of a ring, Lassell showed in a diagram the ring south-preceding, upper left to lower right, thus matching Challis's orientation (see Fig. 4.1).

Lassell had depicted the ring differently, however, in letters to the Reverend William Dawes, a well-known double-star observer. Dawes saw a copy of Challis's March 1847 report.[27] He wrote to Challis on 7 April 1847:

I cannot quite comprehend about the *ring*. Mr. Lassell has not I think distinctly mentioned in any of his published statements respecting it, *in*

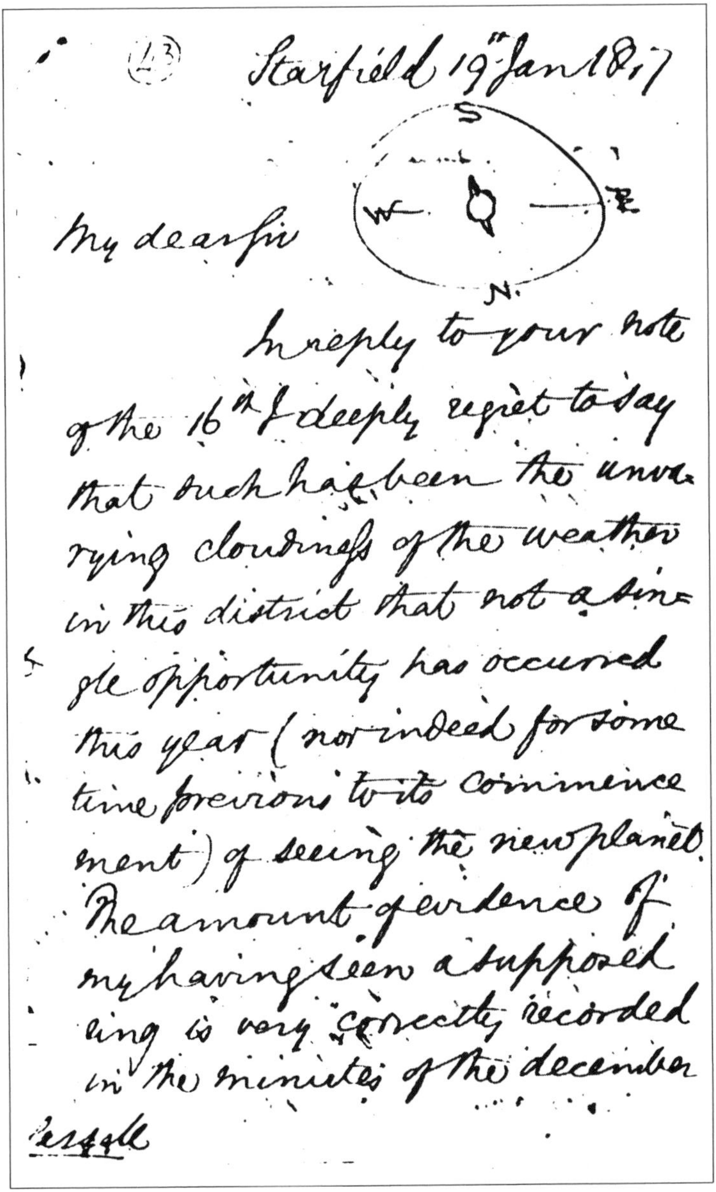

43

Starfield 19th Jan 1847

My dear Sir

In reply to your note of the 16th I deeply regret to say that such has been the unvarying cloudiness of the weather in this district that not a single opportunity has occurred this year (nor indeed for some time previous to its commencement) of seeing the new planet. The amount of evidence of my having seen a supposed ring is very correctly recorded in the minutes of the December

Lassell

Fig. 4.1. Lassell's letter of 19 January 1847 to Challis at Cambridge, depicting Lassell's purported observation of a ring around Neptune. Preconception, in the form of analogy to Saturn with its dramatic ring, may have helped Lassell transform an excess of ellipticity caused by an instrumental defect in his telescope into a full-fledged ring.
Source: Lassell's letter is at the Cambridge Observatories.

what direction the ring lies. He merely says that it is "about 70° from the parallel of declination"; but whether from north following to south preceding, or from south following to north preceding he does not state. — In private letters to myself however he has given me several diagrams from time to time, in which the inclination of the ring has always been drawn from *south* following to *north* preceding; or in double-star phraseology, the angle of position is given 160° and 340°. If this is what Mr. Lassell has observed it is at variance with your observation.[28]

Dawes raised no public question or criticism of Lassell's observations, nor of the apparent discrepancy between Lassell's and Challis's orientations of the supposedly observed ring, but Dawes did pursue the matter to a resolution. In addition to writing to Challis, Dawes apparently also wrote to Lassell and received the reply that Lassell had made a mistake in his letters to Dawes. As Dawes explained to John Herschel:

Lassell informs me that when he drew the position of the ring . . . in his letters to me, the diagram was taken from its appearance in the telescope without respect to the direction of the parallel; — that of the direction being rather referred to. At first he contented himself with noting that it was *nearly at right angles* to the diurnal motion; but in later observations he has expressly depicted it [south preceding to north following]. So that he and Challis agree — Hind draws it [south preceding to north following]; but this of course is of little weight, as well as my own seeing no elongation at all.[29]

Dawes's inability to see a ring with a relatively small 6⅓-inch telescope in bad weather was of little argument against the existence of a ring, as he also stated to Challis.[30] The one time that Dawes had good seeing, he saw Neptune "sharply defined and *circular,* having no projection whatever."[31] Perhaps Dawes had reservations about the claims of Lassell and Challis to have seen a ring, but lacking a telescope comparable in power to Challis's 11¾ inch or Lassell's 24 inch, all Dawes could do was wonder out loud whether Lord Rosse had observed Neptune with his giant 6-foot reflector and if he had seen a ring.[32]

The letter from Dawes to Herschel explaining that Lassell had mistakenly depicted the ring's orientation in his earlier letter to Dawes largely clears up the apparent discrepancy between Challis's and Lassell's observations. It may be that the apparent confusion over the ring's orientation was the result of somewhat sloppy observing by Lassell; he did not initially draw in the direction of motion, notes Robert Smith, and without that fixed reference one could see all sorts of inclination in the telescope, depending on the time of observation.[33] A few nagging doubts, though, remain. It is surprising that Lassell could have made such a mistake. True, he was using a reflecting telescope rather than a refractor; but his Newtonian reflector inverted the image in the same manner as does a refracting telescope.[34] Hence there should not have been any confusion about inverted images. Furthermore, the mistake—if that is what it was—was repeated. In August 1847 Lassell published drawings depicting the ring oriented from upper right to lower left, as he had originally reported it to Dawes—supposedly mistakenly—and the opposite of what he had written to Challis.[35]

What seems to have aroused doubt in Challis's mind about Lassell's observations was not the issue of the general direction of inclination of the ring but Lassell's modification of the angle of inclination. Shortly after receiving Dawes's letter Challis wrote:

With respect to the Ring I have had reason to think that its existence is not so certain a matter. Mr. Lassell in his first observation states the direction of the axis of the Ring to be nearly coincident with a circle of declination [originally, Challis had written "perpendicular to a parallel of declination"]. Subsequently he placed it 20° from this position. And at a still later period I saw it about 25° distant. It has struck me that so good an observer as Lassell could not be so much mistaken as to position in his first observation, and that what we see is something due to the refraction of the atmosphere and consequently changeable in position. It seems nevertheless to be an appearance peculiar to the planet. I am anxious to get other observations, without which I do not now consider the fact of the Ring's existence as decided. I hesitate [Challis changed "hesitate" to "refrain"] on this account to render a drawing of it.[36]

Lassell, who all along had spoken somewhat cautiously of the "supposed ring" while at the same time enthusiastically reporting new confirmations of the existence of the ring, continued in his ambivalent ways. In 1851 he observed "an evident appendage, such as I used to take for a ring," but it vanished at higher magnification.[37] A year later Lassell took his reflector to the clearer skies of Malta, where he had more starlight than in the best seasons in England and quite as much as he had health and strength to use.[38] On the very first night of observing in Malta, on 5 October 1852,

Lassell found Neptune seemingly accompanied by a ring. And in November Lassell wrote in his notebook that "the indication of the supposed ring immediately struck me."[39] The next evening the ring was "remarkably strong" when viewed with a power of 219. The inclination was 80 degrees south-preceding, a shift of 15 degrees from the previous night. With a power of 1018, however, "there was the smallest possible suspicion of the ring together with some elliptic form of the planet—both in the direction of the planet's satellite—query whether not both due to the telescope."[40] Uranus, too, appeared elliptical, thus increasing Lassell's doubts about the reality of his ring vision. On 15 December Lassell finally rejected the ring. He had measured its position "to ascertain whether it depended at all upon the telescope." It did. Rotating the telescope, Lassell found that the ring also moved. "It is thus evident that the phenomenon keeps a constant angle with the direction of the telescope and not at all with the parallel [of declination], proving that whatever may be the cause it is more intimately related to the telescope than the object."[41]

Lassell's observations of a ring were encouraged by an instrumental defect as well as by expectation. He had encountered difficulties in supporting the large mirrors of his telescopes to prevent them from flexing under their own weight. The problem was most severe when the tube was pointed at an object near the horizon, least severe when pointed directly overhead.[42] Lassell observed Neptune near the horizon, as Smith and Baum have noted. They also point to a study of telescope defects by Airy for the light it may throw on Lassell's observations.[43] In 1848 Airy, observing with Lord Rosse's 6-foot reflector, found that while objects seen near the zenith were well defined, the image of a star near the horizon was very defective.

When the eyepiece was thrust in, the image of the star was a well-defined straight line, 20 seconds long, in a certain direction; when the eyepiece was drawn out a certain distance (about half an inch from the former position) the image of the star was a well defined straight line, 20 seconds long, in a direction at right angles to the former. Between these two positions the image was elliptical, or, at the middle position, a circle of 10 seconds diameter. The image of *Saturn* (then without a ring) [its alignment such that the ring was hidden from earth-bound observers] was, in the two positions mentioned above, an oval (not an ellipse), whose length was about double its breadth; or, in the middle position, it was a confused circle, whose diameter was about 30 seconds instead of 20.[44]

Lassell's telescope, too, probably produced an elongated image. Perhaps, Smith and Baum argue, preconception transformed a mere excess of ellipticity into a full fledged ring.[45] Instrumental defect *alone* is not sufficient explanation, because Lassell did not find a similar ring in his test observation of Uranus. Preconception plus instrumental defect seems the culprit in Lassell's case.

For Challis, less excuse is found in instrumental defects. He was observing with a refracting telescope, free of the instrumental defects of a large reflecting telescope, such as Lassell's. Thermal characteristics of a temporary tube used in the initial testing of the Northumberland lens had caused distortion of the image, in one instance causing a star to be intersected by two crossing rays, but Airy designed a permanent tube avoiding this difficulty.[46] An examination of the Northumberland objective in 1937 found it badly striated, but the image in monochrome light was quite circular.[47]

Instrumental defects were less prominent for Challis than for Lassell, preconception and pressure to produce more prominent. Challis, along with Airy, had lost the race to discover Neptune. He had been called upon to make a public accounting of his failure. Lassell, if later encouraged to find satellites and a ring circling Neptune, at least had neither shared in the earlier mandate to search for the planet predicted by Adams and Le Verrier nor lost that race. Nor was Lassell dependent on patronage, other than upon his own personal fortune. Once Lassell with his more powerful telescope announced observation of a ring, an observation seemingly verified by other observers, Challis very nearly "knew" that a ring existed. He had strong incentive to detect it.

If incentive to find a ring was strong, incentive to examine the ring reports critically was weak, at least in any public forum. The discrepancy between Challis's and Lassell's published drawings seems to have escaped public notice. Dawes pursued the discrepancy between Challis's and Lassell's reported ring orientations only in private correspondence and readily accepted Lassell's explanation. Possibly other astronomers, too, noticed the discrepancy and pursued it. Much of the scientific news in England then went via networks, and the nonappearance of an issue in print does not necessarily prove a lack of interest.[48] In the case of Neptune, though, English astronomers had cause for some embarrassment and little cause to stir up further embarrassment, even in the pursuit of scientific truth. Responsibility for exposing and correcting mistaken observations of other observers was diffuse; few felt the responsibility personally.

Astronomers took responsibility for their own observations, at least Lassell did. He was aware of the danger of deception and he exercised precautions. Lassell asked friends unacquainted with his impressions to give their own, and he found their impressions both similar and reassuring. When he thought he saw a ring around Neptune, he checked to see that nearby Uranus was simultaneously free of a similar impression. And when he suspected that the impression of a ring around Neptune was false, he checked and found that Uranus too appeared elliptical. This is a far cry from objectivity, but at least Lassell tried. Also, lest the phenomenon be an artifact of the telescope, he changed mirrors and he rotated the tube. While scarcely forging any advance in the quest for objectivity beyond what William Herschel had made half a century earlier, neither did Lassell regress. Indeed, Lassell's observational notes on a possible Neptunian ring are strongly reminiscent of Herschel's on a possible Uranian ring. Probably Lassell studied Herschel's procedures closely. And, like Herschel's observations, Lassell's too were precariously perched on the edge of objectivity.

5

Planetary Fantasies: Mars

URANUS and Neptune were newly discovered planets, on and about which observers sought to find natural characteristics suggested by analogy with other planets. In the case of the long-known planet Mars, unnatural characteristics were sought, though again, characteristics suggested by analogy.

At the end of the nineteenth century and early into the twentieth, astronomers and the general public became caught up in controversies surrounding the search for intelligent life on earth's neighboring planet. Preeminent among the observers was a newcomer to astronomy, Percival Lowell. His search for signs of human habitation, for artificial canals on Mars, is the most blatant, or forthrightly revealed, instance of preconception motivating and influencing observation. In the controversies that erupted over Lowell's Martian hypothesis are revealed much about the nature of science.

Lowell graduated from Harvard in 1876, took the customary grand tour, though he traveled farther than many, all the way to Syria, and then settled down to work in his grandfather's office. Within six years shrewd investments had freed Lowell from the daily tedium of business. He traveled to the Far East several times, where he was so well received that he was appointed foreign secretary and counsellor for a special Korean diplomatic mission to the United States in 1883. Lowell wrote magazine articles and four books about the region.

Lowell was interested in astronomy. He had made his Harvard commencement speech on the nebular hypothesis, proposed by the French scientist Pierre Laplace in 1796 to explain the origin of the solar system. And on his travels Lowell sometimes took along a telescope. One story has Lowell learning during the course of a trip to Japan in 1892 that the eyesight of the Italian astronomer Giovanni Schiaparelli was failing. Earlier Lowell had been told by a leading ophthalmologist that his eyesight was the keenest the doctor had ever examined; now, so the story goes, Lowell decided that it was his manifest destiny to take up Schiaparelli's observations. By 1892 Schiaparelli's eyesight was deteriorating, but he may not have realized it nor could Lowell consequently have known of it until years later.[1]

During the Martian opposition of 1877, when Mars was nearest earth, Schiaparelli had observed with his 8¾-inch telescope a number of dark, straight lines on the surface of Mars. These he called "canali," meaning "grooves" or "channels." "Canali" was translated into English as "canals," with the implication that intelligent life had constructed the canals. Schiaparelli did not advance theories about the channels, and he was praised by contemporaries for his cautious manner as well as the magnitude of his researches. The historian of science Michael Crowe has argued plausibly, however, that Schiaparelli believed in extraterrestrial life. That belief might have affected his observations of Mars. Pluralist inclinations may also have misled Schiaparelli in his determination of the rotation of Mercury. He asserted that Mercury's rotation period was about 88 days, the same as Mercury's period of revolution around the sun. If so, Mercury would keep the same face always toward the sun, as the moon does toward the earth. The rotation period actually is 58.65 days, two-thirds of the 87.97-day period of revolution. After six rotation periods, approximately equal to four revolutions for Mercury and one for the earth around the sun, Mercury again presents the same face to the sun and is aligned approximately the same with respect to the earth. Thus Mercury sometimes presents features reconcilable with a rotation period equal to its period of revolution around the sun, and Schiaparelli's error, repeated by other astronomers, might thus be explained. As Crowe has pointed out, though, Schiaparelli dismissed some discordant observations, attributing them to a cloudy atmosphere on the planet, and Mercury's purported rotation period was the only one that would spare Mercury's surface alternating frigid darkness and blazing heat. Furthermore, Schiaparelli purportedly found Mercury's axis of rotation so arranged that a fourth of Mercury's surface supposedly saw the sun rise and set every 88 days but never rise directly overhead nor inflict its intense vertical rays on that part of Mercury's surface. Hence, according to Schiaparelli, Mercury had a region of moderate seasons that earthlings might envy. Schiaparelli may have had "brain-directed vision"; the suggestion is plausible, but not proven beyond question.[2]

The prospect of intelligent life elsewhere in the universe was not incompatible with Victorian belief in developmentalism or evolution. Indeed, the climate of opinion positively favored a parallel development of life on earth and Mars.[3] Lowell, in arguing for intelligent life on Mars, was to take "the popular side of the most popular scientific question afloat" and to win a correspondingly large and enthusiastic audience, because, in the words of a critic, "The world at large is anxious for the discovery of intelligent life on *Mars,* and every advocate gets an instant and large audience."[4] Mars and its inhabitants had perhaps developed earlier and thus further than man on earth, in accord with Darwinian theory. Mars revealed the future of the earth, with an advanced science and technology, and also an ebbing of life-sustaining resources. Analogy continued as a source of inspiration and preconception.

Observations of Mars at several oppositions occurring between 1877 and 1892 had added little to Schiaparelli's initial report. Mars was especially close to earth at the 1892 opposition but then could be observed better from the southern hemisphere where few telescopes were located. Nor had a Harvard astronomical expedition to Peru obtained conclusive results. The 1894 Martian opposition would again bring Mars especially close to earth, and Lowell determined to take advantage of the opportunity.

Returning to Boston, Lowell announced in February 1894 that he would establish an observatory in the Arizona Territory and search for signs of intelligent life on Mars. The choice of Arizona in the American desert with its clear and stable air was the suggestion of William Pickering, an astronomer and the younger brother of Edward Pickering, the director of the Harvard College Observatory. In the same month as Lowell's announcement A. E. Douglass, formerly an assistant at Harvard and a member of the expedition to Peru, and now Lowell's assistant, left Boston with a 6-inch telescope to make observations throughout the Arizona Territory and choose the best site for an observatory. Douglass tried several sites in southern Arizona, only to find the visibility there poor. He moved north, soon coming to Flagstaff, a small town in the center of a great plateau in northern Arizona. The visibility was better. The town offered land for an observatory and agreed to build a road to the site. Ground was broken 23 April and regular observations were begun with an 18-inch telescope on 1 June 1894.

Lowell made a series of observations of Mars during 1894 and drew upon the work of other observers at Flagstaff extending into 1895. He promptly published his controversial Martian hypothesis in a series of articles in the *Atlantic Monthly* and in the book *Mars*.[5]

Lowell began his articles and the book with a statement of the probability of extraterrestial life. He cited the essential oneness of the universe and the modesty that forbids man to believe himself the sole thinking being in the universe.

To prove the existence of life on Mars, Lowell asked if the planet was physically habitable and if there were any signs of actual habitation. His criteria of proof were that the hypothesis should not be contradicted by facts and that the probabilities in its favor should be sufficiently great. The point at issue for Lowell was not whether there was a possibility that his theory was false, but whether there was a possibility that it was true.[6]

Much the same argument about the nature of science also is found in a book by Mark Wicks, an English amateur astronomer. In his 1911 science fiction novel *To Mars via the Moon*, Wicks asserted that the refusal to go beyond current knowledge was less scientific than to anticipate new developments intelligently.[7] The argument that something was not proved was to be replaced by the argument that something might be possible; inferences were to be legitimized. The "sane and unsensational astronomy" was to be replaced with more imaginative thought. This more imaginative thought might spread into other areas through the use of the well-established tradition of a voyage to another world as a critique of society at home. Also, science was to be injected with a new enthusiasm.[8]

Lowell's and Wick's philosophy was not shared by professional scientists. A common objection to Lowell's work was that his interpretation of observations, while forming a possible conclusion, was not a necessary conclusion.[9] Henry Norris Russell, the Princeton astronomer, pointed out that there were alternative explanations for the features observed on Mars and that it was therefore necessary to return against the argument for intelligent life the verdict of "not proven."[10] William Campbell at the Lick Observatory stated that those who argued that Mars was inhabited should bear the burden of proof.[11] The issue of speculation and proof divided amateur from professional astronomer in the United States and in Britain.

Speculation quickly led Lowell to a belief in Martian inhabitants.[12] The first step toward a proof of intelligent life on Mars was the demonstration of an atmosphere and water, two things in nature vital for all conceivable forms of life. Lowell reasoned that without an atmosphere all development, including decay, would cease. There were visible changes on Mars: changes in the polar caps, seasonal changes in the tint of the dark areas, and occasional increases in brightness of certain parts of the planet's disc, suggestive of clouds. These changes were conclusive proof of an atmosphere, conclusive at least for Lowell.

Arguing for a considerably thinner atmosphere on Mars than on earth were both the rarity of Martian clouds, in comparison to earthly ones, and Mars' lesser mass, and consequently lesser gravitational hold on its atmosphere. Thin air, however, was not necessarily incapable of supporting a high order of life. Lungs were "not wedded to logic, as public speeches show." Furthermore, fish would undoubtedly argue that life out of water was impossible; to argue that intelligent life was impossible in an atmosphere different from our own was thus to argue not as a philosopher but as a fish.

After air, water also was necessary if Mars were capable of supporting life. For indications of water, Lowell looked to the polar caps and to changes that implied the presence of water as well as air. A dark band appeared around the melting polar cap, in two spots expanding into great bays, blue in color. Light from one of the bays was polarized, as it would be upon reflection from water. Most of the dark areas, however, did not show traces of polarization. The water supply apparently was exceedingly low.

Concluding that Mars did have some atmosphere and some water and thus could support life, Lowell next asked if there were any signs of actual inhabitants. There were signs, visible to Lowell if not to most astronomers. Lowell saw on Mars an amazing network of straight lines, many meeting at the same spot. The lines' straightness, their individually uniform width, and their systematic radiation from central points all helped convince Lowell of their artificial character. On a planet with an apparent dearth of water there appeared an irrigation system of canals carrying water to constructed oases. In the absence of any natural theory explaining the observed phenomena, Lowell concluded that the red planet was inhabited and that man was but a detail in the evolution of the universe.

Most of the facts cited by Lowell were little questioned by astronomers. The polar caps, the seasonal changes in the dark areas, and the occasional bright spots were generally accepted. The point of controversy was Lowell's description of the lines and his interpretation of the lines.

Lowell had the advantage of an effective telescope and, most importantly, "good seeing," steady atmospheric conditions at his high desert observatory.[13] Furthermore, Lowell asserted, prolonged study of Mars enabled him at rare moments of excellent seeing to see much more detail, detail that less practiced observers with less acute eyesight labeled illusory. "Not everybody can see these delicate features at first sight, even when pointed out to them; and to perceive their more minute details takes a trained as well as an acute eye, observing under the best conditions."[14] And Lowell argued, "Experience makes expert, and perception eventually stands secure where it tiptoed at the start."[15] Lowell concluded:

Although to the observer practised in their detection they [fine lines and little gossamer filaments cobwebbing the face of the Martian disc] are at

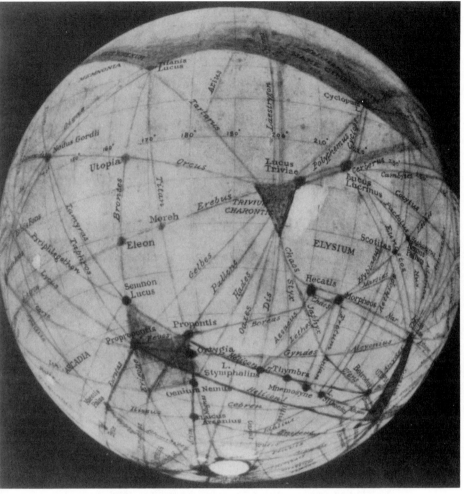

Fig. 5.1. A globe of Mars prepared by Percival Lowell showing the artificial canals he purportedly observed radiating out from oases, carrying precious water on a dying planet. The canals are now attributed to Lowell's imagination rather than to his telescope, fine observing conditions, and extraordinary eyesight.

Source: The photograph of the globe of Mars is reproduced by permission of the University of Arizona Press from its edition of William Graves Hoyt, *Lowell and Mars* (Tuscon: Univ. of Arizona Press, 1976) and by permission of the Lowell Observatory. "Lowell Observatory Photograph."

certain times not only perfectly distinct, but are not even difficult objects, — being by no means at the limit of vision, as is often stated from ignorance, — to one not used to the subject, and observing under the average conditions of our troublesome air [despite Lowell's frequent assertion of excellent seeing, nearby mountain peaks caused eddies in the air and the average seeing at Flagstaff was considerably less good than rare moments of excellent seeing], they are not at first so easy to descry. Had they been so very facile, they had not escaped detection so long, nor needed Schiaparelli, the best observer of his day, to discover them. But in good air they stand out at times with startling abruptness. I say this after having had twelve years' experience in the subject — almost entitling one to an opinion equal to that of critics who have had none at all.[16]

Not for Lowell the humility of a William Herschel or a William Lassell, doubting their vision and fearful of deception. If others could not see what Lowell saw, it was because they could not see as well as he did. They had neither acute eyesight, nor years of experience, nor the best observing site. Arrogantly confident, not without partial justification, Lowell proceeded toward his manifest destiny with scarcely a backward look. Precautions that Herschel and Lassell incorporated into their observations were foreign to Lowell's spirit.

Douglass, less blinded by Lowell's spirit, developed doubts about some of the Martian observations. He could see single canals on Mars in profusion and did not question their reality. Douglass could also see double canals, two lines running side by side. Schiaparelli had noticed a few double canals, and Lowell and Douglass added to the number observed. Douglass, however, doubted the reality of the double canals, especially because of an apparent relation between the widths of the double canals and the aperture of the telescope.[17] Here, seemingly, was an artifact of the observing tool. Douglass wrote:

My opinion is, that they are right who attribute duplication of the canals chiefly to a subjective effect; I am inclined to think that the cause is not poor focusing, but a misinterpretation of crowded canals [mistaking two adjacent and slightly divergent canals for one double]. I have drawn double canals a number of times, but nearly always have expressed in the notes at the same time some reservation about them. Sometimes this reservation is a statement that I would not have suspected the duplication unless told of it. Another time it was that I would have suspected the duplication even if it had been unheard of. Again, and only once, I said "Positively see this double, but not all the time."[18]

Lowell's opinion was that those who questioned the duplication of the canals were wrong. Perhaps Lowell had early entertained some doubts; Douglass did comment that Lowell was inclined to agree with the criticism of the double canal observations.[19] But if Lowell ever had doubts, they soon vanished. So did Douglass, whom Lowell abruptly discharged after seven years of devoted service, for "untrustworthiness."[20]

The decision that Douglass was untrustworthy apparently was precipitated by a letter from Douglass to William Putnam, Lowell's brother-in-law and supervisor of the observatory during Lowell's nervous exhaustion between 1897 and 1901. Douglass complained about Lowell's overhasty publication of sweeping conclusions on the basis of few and selected observations. "His method is not the scientific method and much of what he has written has done him harm rather than good. I fear it will not be possible to turn him into a scientific man."[21]

Douglass had begun an investigation, against Lowell's wishes, to determine whether disputed planetary details might be psychological in origin. To a psychologist Douglass confided, "I would have written you long before but for Mr. Lowell's indifference to taking up the psychological question involved in astronomical work." Notwithstanding Lowell's indifference, Douglass had "made some experiments myself bearing on these questions by means of artificial planets which I have placed at a distance of nearly a mile from the telescope and observed as if they were really planets. I found at once that some well known planetary appearances could, in part at least, be regarded as very doubtful. . . ."[22]

Little came of Douglass' 1901 study, but Lowell himself made observational and psychological studies in 1903 to counter criticisms. The investigations seem intended primarily for polemical effect.

Douglass had mentioned an apparent relation between the width of double canals and the telescope aperture, indicating an instrumental effect. He had thought it was a case of mistaking two canals for one double, which would happen only near the limit of separating power of the telescope; thus the width of the mistaken double canal would have a dependence on the size of the instrument. Another possible explanation of the phenomenon of apparent double canals lay in optical theory. Light waves from a bright line after refraction by a telescope produce alternating dark and light bands on either side of the image and thus could produce the appearance of a double line. The width of such a double line would vary with the aperture of the telescope. Lowell quickly dismissed the notion that the double canals were not real, regardless of what explanation was offered, by observing, so he said, the same width with different apertures.[23] Lowell did not acknowledge that other observers had asserted a dependence of double canal width on telescope aperture.

Another criticism of Lowell's observations concerned how wide a canal on Mars would have to be before it would be visible as far away as the earth. Optical theory suggested a width of many miles. Lowell's experiment, consisting of stretching a telegraph wire of known width from the telescope dome and then determining from how far away he could see it, suggested that a canal as little as 3/16 of a mile wide would be visible at Mars' closest approach to earth. The difference between theory and experience was explained in psychological terms.

Why a line can be seen when its width is but 1/86 of the *minimum visible* [according to optical theory] seems to be due to summation of sensations. What would be far too minute an effect upon any one retinal rod to produce an impression becomes quite recognizable in consciousness when many in a row are similarly excited. Psychologically it is of interest to note that there are stimuli perceptible so faint and so fleeting as to be even below this limit, and that unable to rise into direct consciousness, leave only an indefinite subconsciousness of their presence which the brain is unable to part from its own internal reverberations. It is a narrow limbo, this twilight of doubt. . . .[24]

Even Lowell was now impressed with how precariously observations were perched on the edge of objectivity, at how narrow was the border or space between reality and illusion, and he directed a repetition of his telegraph wire experiment in which, "as a check against any influence that a knowledge of the positions of the wire and line might introduce, the observer V. M. S. [Vesto M. Slipher] had nothing to do with the preparation and arrangement of the experiment . . . results are practically the same for the two observers—one having no knowledge as to the position of the objects. . . ."[25]

The major challenge to Lowell's observations came not from

optical theory regarding the minimum visible width of a canal on Mars, though, nor from the possibility of interference creating bands to be confused as doubled lines. Presenting Lowell with his greatest challenge, and doing most to sway opinion away from Lowell's Martian hypothesis, was the demonstrated tendency of the human eye to connect well-seen points by imaginary lines.

Walter Maunder, the photographic and spectroscopic assistant at the Royal Observatory, asked a group of boys from the Royal Hospital School at Greenwich to copy a picture of Mars from which Maunder had removed all traces of canals. In copying the picture, the boys added the lines back in. Maunder concluded that the apparent lines on Mars were "simply the integration by the eye of minute details too small to be separately and distinctly defined."[26]

To combat the "small boy theory" of optical illusion, Lowell devised and performed a series of experiments. He systematically observed telegraph wires of various thicknesses at various distances. Generally, Lowell and other observers were able to distinguish real wires from illusory ones. Illusory lines did appear at great distances and near the limit of visibility. There was a tendency for the human eye to connect real points with imaginary lines, but "the whole art of the observer consists in distinguishing which of these phenomena are objective and which are not. . . . A little more experience than the boys possessed would have permitted of parting the true lines from the false. . . . Because a small boy would certainly not distinguish Oolong from Soochong teas by their taste, it were unsafe to assert that a professional tea-taster cannot."[27]

Simon Newcomb, an astronomer and the most honored American scientist of his lifetime, repeated Maunder's experiment with astronomers from the Harvard and Yerkes observatories. Newcomb accepted "the general principle that what is seen by a single practiced observer under the most favorable conditions affords evidence which completely outweighs that of less favored observers," but not if there was absolute inconsistency between observers, as there was between Lowell and many astronomers. Newcomb also accepted Lowell's criticism that Maunder had used inexperienced observers, the school boys, and Newcomb substituted experienced observers. They, too, found that at increased distances discontinuous objects were perceived as lines."[28]

Newcomb investigated both optical and psychological causes

affecting the judgment of an observer scrutinizing faint and difficult features on the surface of a planet. Under the psychology of
vision, Newcomb recognized two distinct processes, the conscious
stimulus of optical nerves by light and the perception by the mind
of a real or supposed object indicated by such stimulus. He used
the term visual inference "to describe the act by which the mind
unconsciously draws conclusions as to an observed object from the
image formed by its light on the retina. A fundamental property of
this form of influence is that it comprises not only *seeing,* in the
ordinary sense, but a rational interpretation or conclusion based on
previous experiences of what is seen. . . . In this process we have a
possible fruitful source of error of vision which, instead of being
corrected by experience, tends to be strengthened by it. A mind
accustomed to dealing with objects the correct perception of which
depends mainly on visual inference, is naturally prone to extend
that inference to cases where the conclusion would be illusory. Having this in mind, we see that observers trained in different ways
may depict the same object very differently."[29]

Lowell responded to Newcomb's article, particularly the discussion of optical causes, with an impressive mathematical attack.
Lowell had been such an outstanding mathematics student at Harvard that his professor had hinted that if Lowell applied himself to
the subject he could hope to succeed his teacher to the chair. To
Newcomb's psychological discussion, Lowell replied that Newcomb's experimental disc, upon which observers tended to connect
discontinuous objects, was less than a tenth of the size of Mars'
image in a telescope; to be comparable to the image of Mars,
Newcomb's disc had to be viewed from a distance of only 9 feet,
and then there were no illusions.[30] A note by Newcomb respond-
ing to Lowell's paper and a note by Lowell replying to Newcomb's
note ended the dispute rather inconclusively.[31]

Lowell's tea taster analogy may have been telling, as was much
of his witty repartee; but Maunder's small boy theory was more
telling, even before Newcomb using experienced observers duplicated Maunder's result. Privately, Lowell had claimed support
against Maunder's charge from Camille Flammarion, who had
worked at the Paris Observatory before becoming a major popular-

izer of astronomy. According to Lowell, Flammarion repeated Maunder's drawing experiment with French school boys and not one of the boys drew in an illusory line.[32] Therefore, Lowell asserted, the English astronomer Sir Norman Lockyer concluded that Maunder had put a leading question to his group of boys.[33] Whatever solace may have been offered Lowell in private correspondence, however, few astronomers came to his defense in public.

Lowell's unproven leaps from possibility to stated hypothesis were objectionable to astronomers. Even more objectionable in Lowell's work was the element of preconceived opinion publicly stated. Campbell at the Lick Observatory wrote in his review of Lowell's book *Mars*: "Mr. Lowell went direct from the lecture hall to his observatory, and how well his observations established his pre-observational views is told in his book."[34] And in his 1916 obituary of Lowell, Russell warned:

if the observer knows in advance what to expect . . . his judgment of the facts before his eyes will be warped by this knowledge, no matter how faithfully he may try to clear his mind of all prejudice. The preconceived opinion unconsciously, whether he will or not, influences the very report of his senses, and to secure trustworthy observations, it has been recognized everywhere, and for many years, he must keep himself in ignorance of what he might expect to see.[35]

In beginning his observations with a preconceived opinion, publicly stated, Lowell had violated accepted scientific procedure.

Lowell's critics seemingly also violated accepted practice in criticizing Lowell openly. Russell's obituary notice was, in the circumstances, naturally somewhat restrained in its criticism of Lowell, and all the more unusual for containing any criticism at all. Campbell's book review, if scathing in parts, at least stopped short of character assassination. Other critics were not so circumscribed. Eliot Blackwelder, a professor of geology at the University of Wisconsin, took exception to Lowell's 1908 book *Mars as the Abode of Life*. In choosing Laplace's nebular hypothesis, Lowell had slighted the rival Chamberlin-Moulton hypothesis on the formation of the solar system. Thomas Chamberlin had been president of the University of Wisconsin before going on to the University of Chicago. Blackwelder characterized Lowell's book as fancy foisted upon a trusting public. Misbranding intellectual products was as immoral as misbranding manufactured products, and "censure can hardly be too severe upon a man who so unscrupulously deceives the educated public, merely in order to gain a certain notoriety and a brief, but undeserved, credence for his pet theories."[36] The quarrel escalated in an entertaining if unseemly manner, and Forest Moulton contributed a description of Lowell as "that mysterious 'watcher of the stars' whose scientific theories, like Poe's vision of the raven, 'have taken shape at midnight.' "[37]

The restraint observed by potential critics in earlier instances of preconception and failed objectivity was abandoned in the Lowell case. Lowell claimed but weakly sympathy from astronomers and scarcely any collegial feeling at all from geologists. Lowell was not accepted as a professional scientist by professional scientists. Marginally, he might be an astronomer; certainly he was not a geologist. In criticizing Lowell, professional scientists were disassociating themselves from a failure of objectivity, not revealing their own failure in the quest for objectivity. They were cutting out a cancer before it spread, in the public's mind, to science.

Astronomers, holding the ideals of observation without prejudice of preconception and deduction rather than speculation, responded negatively to Lowell's writing. Other readers had different values, and their responses to Lowell were often quite favorable. For cultured New England readers with demanding literary standards, Lowell wrote entertaining and informative prose. He magnificently and eloquently described Mars as "a great red star that rises at sunset through the haze about the eastern horizon, and then, mounting higher with the deepening night, blazes forth against the dark background of space with a splendor that outshines Sirius and rivals the giant Jupiter himself."[38]

Readers responded enthusiastically to the literary merits of Lowell's writing. He was seen as something of a poet turned physical scientist, who applied the New England intellectual heritage to a previously desiccated science. In literary circles Lowell was accepted as the highest living authority on the subject of Mars and America's most eminent living astronomer.[39]

Lowell's more than competent style no doubt increased acceptance of his theory. His writings also fell on fertile ground. Translations and summaries of popular articles by both Schiaparelli and Flammarion had already appeared in the United States, creating considerable public interest in Mars and in the possibility of extraterrestrial life. Cynics attributed the increasing interest in Mars to a combination of journalistic enterprise and public imbecility.[40]

Further fueling the enthusiastic response to Lowell's Martian hypothesis were complementary beliefs. The idea of tenantless globes revolted many a man's sense of the rational in creation. Discovery of intelligent life on other planets would increase reverence for the Creator. There would also be enlightening social results. Cooperation was the distinctive feature of life on Mars; earth dwellers might benefit from this example of a civilization that did not have wars.[41]

Predisposition to find signs of intelligent life on Mars thrived. Scientific, evolutionary theories suggested parallels between the development of life on earth and Mars. Various social beliefs, values, and ideologies drew upon observations of Mars for support through analogy or example. Responses from both the scientific community and from other sectors of society to Lowell's vision of canals on Mars indicate some of the myriad of means by which suggestion and preconception may intrude themselves into otherwise objective scientific observation.

Available in the quest for objectivity, had anyone wished to use them, were observational procedures developed by Herschel, Lassell, and others. They had recognized psychological threats to objectivity and had taken precautions. While far from perfect, the precautions followed were at least an attempt to achieve objective observation. In the case of Lowell and canals on Mars, however, interest focused more on polemical battles between opponents and less on a disinterested search for truth. Douglass in his investigation of psychological effects is an exemplary exception, as well as a casualty.

Without the sheltering cloak of deference normally accorded a scientific colleague, Lowell was an unusually attractive target for criticism publicly expressed. He was open to criticism for not devising and exercising precautions against preconception. Instead of reasonably criticizing him for this failure, however, American astronomers criticized Lowell for possessing preconceptions and for revealing them in public.

The implicit assumption that scientists could study nature free of preconceptions is not consonant with human nature. In focusing on the undesirability of preconception rather than on how to reduce the deleterious effect of inevitable preconception, American astronomers regressed in the quest for objectivity. In overlooking the hard-won results of English predecessors, American scientists may have increased the likelihood of future embarrassments, of future failures of objectivity.

6
Sirius B and the
Gravitational Redshift

INSTANCES of failed objectivity in the history of science might perhaps be dismissed as past history, of little relevance to the practice of contemporary science. No historian would so cavalierly cut off current science from its roots. Historians, though, are among the first to seek characteristics exemplifying the evolution of science, characteristics defining contemporary science in ways different from the science of previous centuries.

Harriot, Hooke, Flamsteed, and other seventeenth-century scientists expressed little conscious awareness of the problem of personal bias and its effect on scientific observation. They developed none of the checks necessary for its control.

William Herschel at the end of the eighteenth century learned firsthand with his observations of the diameter of Uranus of the problem posed by expectation. His subsequent work constitutes a major step in the quest for objectivity in scientific observation. Herschel, later Lassell in his observations of Neptune, and later yet Lowell observing Mars made test observations of comparison objects in commendable if not always successful attempts to eliminate the possibility of instrumental defect, optical illusion, and personal bias.

Still, Herschel, Lassell, and Lowell were scarcely professional scientists employing all modern precautions to assure objectivity. An examination of more contemporary, twentieth-century instances of failed objectivity among the best professional scientists at the best scientific institutions is crucial to any study of objectivity and modern science. Three such instances are known to have occurred at the Mount Wilson Observatory. The first is the case of Sirius B and the gravitational redshift.

Sirius, one of the brightest stars in the sky, was long and often observed. Not until the nineteenth century, though, would its double nature be discovered — the faint companion Sirius B resolved from the much larger Sirius, now Sirius A. The German astronomer Friedrich Bessel, one of the most skillful observers of his century, in 1834 noticed irregularities in the motion of Sirius. A decade later Bessel suggested that the irregular motion was caused by the gravitational force of an invisible companion.[1] During the following two decades astronomers would make more detailed calculations of the motion and force.[2]

The first actual sighting of the predicted but elusive faint companion occurred by chance in 1862. Alvan Graham Clark, one of the sons in the American optical firm of Alvan Clark and Sons, saw Sirius B in the course of testing the lens for a new 18½-inch telescope, the first of a series of five world's largest refracting telescopes to be built by the Clarks over the next thirty-five years.[3] The serendipitous discovery of Sirius B, for the Clarks probably had not known of the theoretical prediction, was soon confirmed.[4]

In 1890 Sirius B was lost to view, passing behind Sirius. In 1896 the American astronomer T. J. J. See, then at the Lowell Observatory, claimed the rediscovery of Sirius B. But See's alleged position for Sirius B was not where other astronomers were soon to find it. Here is a possible case of preconception producing observation, but scarce facts do not support a definitive explanation. And given See's character, outright fraud seems as plausible an explanation as failed objectivity.[5]

Sirius B's luminosity was mysteriously low, relative to its normal mass, and interest in Sirius B was thus high. Especially wanted was a spectrum of Sirius B, its light spread into components by a prism or a grating, for the hints it might offer about the star's physical nature.

The low luminosity helped sustain the mystery, because light from adjacent Sirius, ten-thousand times as bright as Sirius B, made it virtually impossible to obtain a separate spectrum of the fainter object. In 1915, however, success was reported by Walter Adams, an experienced observer at the Mount Wilson Observatory, using the 60-inch reflecting telescope, then the world's largest. Adams's surprising and important discovery, whether valid observation or fortuitous coincidence, was that the faint companion is not a relatively cool red star but is white hot, as is Sirius.[6]

A white-hot star of normal mass with extraordinarily low luminosity soon would be brought within the context of relativity theory, just then being developed. From relativity theory Einstein had predicted three astronomical phenomena: the motion of Mercury in the sun's gravitational field; the bending of light in a gravitational field; and a gravitational redshift, a small shift of light waves to longer or redder wavelengths in the presence of a very strong gravitational field.[7]

Already, in 1912, Einstein could cite in tenuous support of his prediction of a gravitational redshift several observations of displacements in solar spectra, displacements possibly caused by the sun's gravitational field.[8] Next, Einstein enlisted Erwin Freundlich at the Royal Observatory in Berlin to check the three relativistic predictions.[9] Freundlich worked first on Mercury's motion. Then in 1914 he set off on a solar eclipse expedition to the Crimea, hoping to detect the bending of starlight as it passed near the sun. What he found was the outbreak of World War I, and Freundlich was soon a prisoner of war, subsequently exchanged for captured Russian officers.[10]

In the meantime, astronomers at the Kodaikanal Observatory in India had made further observations of displacements in solar spectra. Then unaware of Einstein's theory, they interpreted the displacements as motions of solar material or density changes.[11] Freundlich, however, was quick to seize upon the possible significance of the measurements for Einstein's theory.[12]

The initial measurements from India agreed very closely with Freundlich's theoretical calculation, but further observations did not.[13] They were distinctly unfavorable to Einstein's theory.[14] Nor were astronomers in Germany and in the United States successful in their efforts just before, during, and immediately after the war to detect a gravitational redshift in the spectrum of the sun.[15] Arthur

Eddington, director of the Cambridge Observatory, measurer in 1919 of the bending of light in the sun's gravitational field, and one of the strongest advocates of relativity theory, was reduced to pleading for a reservation of judgment on the question of a gravitational redshift.[16]

Eddington came to believe that white dwarf stars, such as Sirius B, are very dense. He sought to verify his prediction of a high density for white dwarf stars with the detection of a gravitational redshift for Sirius B. While the sun's gravitational redshift was seemingly too small to detect, Sirius B's might be measurable. The gravitational redshift is proportional to the density, and Sirius B's density was estimated by Eddington to be a phenomenal ton per cubic inch.[17] It would be necessary to sort out any Doppler redshift component due to motion, readily done in the case of double stars by taking the difference between their motions. Hence Sirius B was an ideal test case for Eddington's white-dwarf theory.

It was also an ideal test for Einstein's relativity theory and the prediction of a gravitational redshift. Perhaps this fundamental goal had been lost sight of by Eddington in his enthusiasm for his own white-dwarf theory. Or, convinced of the basic truth of relativity theory, Eddington may have naturally focused on the more problematical issue, especially after his English rival James Jeans challenged Eddington's theory.[18] Only the world's largest telescope would suffice for the test, and Eddington asked Adams at Mount Wilson to attempt the measurement.[19]

No doubt Adams appreciated that some fame would attach to the first astronomer to measure a gravitational redshift. Adams may also have felt pressure to produce because he had the best instrumentation then available and thus the best possibility of success. Perhaps, also, production would justify the past expense of his instruments and help in the acquisition of further funding. It may well be that professional pressures, with their attendant danger of a heightening of the effect of preconception, focus most sharply on the best scientists and the best scientific institutions.

Adams recognized the opportunity to complete Einstein's third test of relativity theory. He believed that there existed a gravitational redshift to find. He knew what value was expected for the redshift.

Actually, the expected value changed before Adams was through. Early in 1924, when Eddington first wrote to Adams, the best prediction was a redshift of 28.5 kilometers per second.[20] From plates already obtained, but set aside because Adams had not before realized how large the redshift might be, he measured a simple mean of 20 kilometers per second. If the best spectral line were given double weight, though, the mean became 23 kilometers per second, closer to Eddington's predicted value. And, of course, there was a large probable error, so observation was not incompatible with theory.[21]

Eddington responded, lowering his calculated value. He had believed Sirius B to have a type A spectrum, but the F0 spectrum observed at Mount Wilson meant a lower temperature and a lower redshift, of 19 kilometers per second. But the temperature might be a little larger in this type of star to create an F0 spectrum, and Eddington concluded that 20 to 25 kilometers per second was the most likely value.[22]

Adams obtained more spectra for what he believed to be Sirius B, not the much brighter Sirius. His photographs showed both hydrogen lines and lines of some metals. Obviously Adams's spectra, if of Sirius B at all, were seriously contaminated with light from Sirius, since metallic lines do not occur in the spectra of white dwarf stars. But this fact was not to be known until much later, and Adams blithely proceeded to measure displacements ranging from plus 33 to minus 2 kilometers per second. Application to the data of several somewhat arbitrary weighting factors produced an average differential redshift of first 21, and later 19 kilometers per second.[23]

Eddington was delighted with Adams's result.[24] But Eddington was also cautious. He wrote: "however experienced the observer, I do not think we ought to put implicit trust in a result which strains his skill to the utmost until it has been verified by others working independently.[25] For Eddington personally, Einstein's theory gave stronger assurance of the existence of a gravitational redshift than did the observational evidence.[26] Of course, Einstein's relativity theory had by this time received considerable backing from observation, most notably Eddington's own detection of the bending of light in a gravitational field. Thus Eddington's stated preference to rely on Einstein's theory rather than on Adams's observation does not necessarily represent a complete disdain for observation. It may be that Eddington was suspicious of Adams's purported measurement, a suspicion well justified by subsequent results.

Somewhat ironically, as the English astronomer William McCrea has noted, the Royal Society seized upon Adams's measurement as reason to award its medal to Einstein. The society had missed the earlier opportunity presented by Eddington's detection of the bending of light, perhaps because the result had not yet been published and thus subjected to scrutiny at the time when the vote on the award of the medal had taken place.[27] Actually, Einstein was fortunate that the bending of light had not been measured earlier. McCrea asks how Einstein's relativity theory might have been received had Freundlich succeeded in measuring the bending of light in 1914, when Einstein's calculated value was too low by nearly half, thus denying Einstein the opportunity to correct his calculation prior to the first measurement by Eddington in 1919.[28] No doubt Einstein's theory subsequently corrected would have had an unattractive and unconvincing appearance, more ad hoc than fundamental.

The Royal Society also had missed the opportunity to recognize and honor Einstein presented by Einstein's prediction of the motion of the perihelion of Mercury, an irregularity in the motion of Mercury's orbit unaccounted for by Newtonian gravitational theory.[29] This prediction, or retrodiction, had followed rather than preceded observation, and there is a considerable psychological if not strictly logical difference between the relative values of prediction and retrodiction in confirming a theory. The prediction of phenomena in advance of their observation lends conviction to the theory producing the prediction. Such conviction should be tempered, however, by knowledge of instances in which prediction has led to observations of phenomena now known not to exist.

Eddington had recommended caution until Adams's observation was independently verified. Apparent confirmation from an independent observer, independent except for the shared expectation of finding a differential redshift of about 19 kilometers per second, seemingly was forthcoming. In 1928 Joseph Moore, whose main task at the Lick Observatory was to measure redshifts in the spectra of stars, reported that he had measured differences between the redshifts of Sirius and Sirius B of 5, 16, 17, and 24 kilometers per second. After excluding the low value because it was seriously affected by the superposition of scattered light from Sirius, Moore came up with an average value of 19 kilometers per second, precisely what Adams had reported earlier.[30]

Notwithstanding Moore's purported confirmation, Adams's alleged measurement of the gravitational redshift of Sirius B is no longer accepted. The measurement had agreed nicely with theory in 1925, but theoretical understanding of white dwarf stars was later modified and by 1950 a disparity had risen between theory and Adams's measurement.[31] In the late 1960s a new measurement of the gravitational redshift of Sirius B completely discredited the earlier work. Astronomers at the Mount Wilson Observatory found a gravitational redshift for Sirius B of over 80 kilometers per second, with an estimated error of not more than 16 kilometers per second.[32] The new measurement is far too large to be reconciled with the earlier value of only 19 kilometers per second. The new, higher value is retrodicted easily enough if the radius of Sirius B is revised downwards and the density correspondingly increased.[33] Unfortunately, the new measurement cannot be easily verified because it was made under the unusually good condition of newly coated telescope mirrors.

Recognizing the incompatibility of his new measurements with the old, Jesse Greenstein at Mount Wilson attempted to explain away the latter. He stated that Adams's and Moore's spectra were badly contaminated and that their results depended upon metallic lines now known not to occur in white dwarf stars. Hence the measurements are "of historical interest only," according to Greenstein.[34] But what historical interest! Adams's result cannot be completely blamed on contamination by metallic lines. He found the metallic lines so faint in the spectrum of Sirius B that he based his result primarily upon measurements of two hydrogen lines. Adams traced intensity curves of the hydrogen lines with a microphotometer and he also measured the hydrogen lines with a comparator. Not only did Adams find a differential redshift from the hydrogen lines, but the wavelength displacement was greater for the H_β than for the H_γ lines, as expected, since the wavelength displacement is proportional to the wavelength of the spectral line. Furthermore, even had Adams's result been based entirely upon metallic lines, there would still be the problem of how Adams could have measured a differential redshift in spectral lines from the same star.

Here is a modern instance of professional scientists, Adams at Mount Wilson and Moore at Lick, eluding the constraints of objectivity to find what they expected to find, even when it didn't exist. Here also is an instance of scientists later accepting uncritically, at least without objection in print, a patently inadequate effort to explain away in the context of instrumental error rather than failed objectivity an otherwise embarrassing incident.

There is something to be said in favor of silence, however little the facts justify acquiescence in the metallic-line-contamination explanation. The earlier observations cannot be explained away in an acceptable manner, excluding the possibility of failed objectivity. If all suggestion of failed objectivity is taboo, potential discussion is terminated. Nor is there much of a problem in choosing between the rival observations. To the extent that a more powerful telescope promises better observations, the more modern measurement is to be preferred, given a choice between two otherwise equal but conflicting claims.

Little, other than historical insight, accuracy, and integrity, is to be gained by a critical examination of explanations of Adams's and Moore's errors. Much might be lost, particularly the prestige and patronage enjoyed by modern science and scientists. A strategy of selective myopia or selective ignorance has practical appeal.

7

A General Solar Magnetic Field

THE SECOND of three known instances of failed objectivity in twentieth-century astronomy is the purported measurement of a general solar magnetic field. Unlike the measurement of the gravitational redshift of Sirius B, largely the lone endeavor of a sole astronomer and performed without elaborate checks and precautions to block the bias of preconception, the solar measurements were the product of several Mount Wilson astronomers consciously aware of the problem of achieving objectivity and carefully pursuing this estimable goal. But not carefully enough.

George Ellery Hale, the founder and director of the Mount Wilson Observatory, led the project to measure a general solar magnetic field. Among the astronomers performing measurements on photographic plates taken by Hale was the young Dutch astronomer Adriaan van Maanen. From his apparent solar success, van Maanen would go on to create, largely by himself, an even more notorious instance of failed objectivity.

As a youth Hale's greatest ambition had been to photograph the sun's spectrum. He succeeded around 1884, while still in his teens, using a telescope and small prism spectroscope provided by his father.[1] Father, whose wealth from the manufacture of elevators rose with the new buildings of Chicago, kept young George provided with scientific books and instruments. A complete spectroscopic laboratory, the Kenwood Observatory, ultimately came into existence with George its master.

Elevators paid for the small Kenwood Observatory; streetcars soon paid for a grander establishment with larger instruments. Returning to Chicago after graduation from the Massachusetts Institute of Technology, where he had majored in physics and written his senior thesis on the photography of solar prominences, Hale added to the instruments and his research results at the Kenwood Observatory and taught astrophysics at the University of Chicago. Upon hearing in 1892 that two 40-inch telescope lenses were available for sale because a Southern California land boom had burst, Hale persuaded Chicago streetcar developer Charles Yerkes to buy the lenses and build an observatory. The Yerkes Observatory was completed in 1897.

Meanwhile, Hale had persuaded his father in 1896 to buy a reflecting disc for an even larger telescope. The University of Chicago, however, couldn't raise the money to mount it. Not until 1902 were the necessary funds forthcoming, from the Carnegie Institution of Washington. Hale then founded the Mount Wilson Solar Observatory on a peak above Los Angeles. In 1908 Hale began observing with the 60-inch telescope, then the largest in the world.[2]

Four years later van Maanen joined the Mount Wilson staff on the recommendation of J. C. Kapteyn. At Groningen, in the Netherlands, Kapteyn had turned to advantage the apparent liability of not possessing a telescope. Partially with convicts from the state prison placed at his disposal, Kapteyn skillfully carried out the tedious but necessary task of measuring photographic plates. The Astronomical Laboratory's first major achievement was a thirteen-year project measuring the position and brightness of some half a million stars on plates of the southern skies taken at the Royal Observatory in Cape Town, South Africa. Next came participation in an international effort measuring the brightness and motions of stars.[3] It was near the midpoint of the international project, from 1908 to 1910, that van Maanen worked under Kapteyn at Groningen, measuring the motions of over 1,400 stars. This was the basis for van Maanen's doctoral dissertation at the University of Utrecht, completed in 1911.[4] Next, van Maanen visited the United States and served as a volunteer assistant at the Yerkes Observatory. Kapteyn customarily spent his summers as a research associate at the Mount Wilson Observatory, and in 1912 he recommended van Maanen for a position there.[5] Thus van Maanen arrived at Mount Wilson, an ocean and a continent away from his scientific training.

Skills developed at Groningen would soon be employed by van Maanen to measure motions of stars and nebulae, but one of his first tasks at Mount Wilson was to help Hale measure the strength of the sun's general magnetic field. The solar measurements are important for the question of personal bias and objectivity in science. At the time of his death in 1946, astronomers regarded as unfortunate any shadow on van Maanen's long series of difficult measures of the sun's magnetic field cast by the later error in his work on spiral nebulae.[6] More recent opinion sees similarities in the two cases.[7] Also, some of the instruments and procedures used in the solar work would be adapted for the subsequent measures of spiral nebulae.

The purported measurement at the Mount Wilson Observatory of a general solar magnetic field followed from newly designed instruments, several converging lines of research, and a succumbing to anticipation or expectation based on physical theories. One of the new instruments was a heliomicrometer especially designed to measure the position on a spectroscopic plate of spots of changeable form. Techniques developed for measuring positions of solar markings later would be used by van Maanen to measure positions of knots in spiral nebulae, which led to his spurious assertion of the rapid rotation of spiral nebulae. The few sunspots on a photographic plate were easily measured and their positions determined, but flocculi presented a difficulty. These bright regions of the sun, now called plages, numbered forty or fifty per plate (approximately the number of knots later measured by van Maanen in spiral nebulae) and underwent rapid changes of form, making their centers and consequently their positions difficult to define precisely (as difficult to define as would be the knots in spiral nebulae). Hale modified his heliomicrometer, mounting two telescopes parallel to each other. One pointed at a photograph of the sun and the other pointed at a globe with marked reference lines. The two images were brought together in a single eyepiece and the positions of the flocculi then read off from the reference lines. Moveable cross hairs added further to the accuracy of the position measures. Thus modified, Hale noted, the heliomicrometer could be used as a stereo-comparator simply by substituting a second photograph for the globe. And he used it thus, to compare the forms of flocculi photographed on successive days and the motions of the flocculi over time, a measure of solar rotation.[8]

Measuring the rotation of the sun was but a minor part of Hale's research. Of greater interest were sunspots, especially their connection with magnetic fields.[9] Many elements are ionized, or become electrically charged, at the high temperatures existing in the sun, and a rapidly rotating electrically charged disc had been shown to produce a magnetic field.[10] Hale, upon discovering vortices in sunspots, reasonably concluded that sunspots should have magnetic fields.[11] A means of testing this hypothesis was at hand. Under laboratory conditions, spectral lines had been seen to widen and even split into doublets in the presence of an intense magnetic field.[12] Visual observations of sunspot spectra over several decades had detected a widening of some lines, a fact confirmed photographically.[13] From his examination of photographs Hale concluded that magnetic fields probably existed in sunspots.[14] Pieter Zeeman, the Dutch physicist who had discovered the Zeeman effect of magnetism on radiation, agreed.[15]

From sunspots Hale turned his search for magnetism to the sun as a whole. It is a rotating body. Furthermore, polar plumes in the corona, the sun's outer atmosphere, had been observed to resemble in shape the field lines of a magnetized sphere.[16] The sun's large size and velocity were expected to produce a large magnetic field, easily detected by Zeeman splitting in spectra of light from luminous vapors in the sun's atmosphere. The intensity of the field, and thus the amount of observed line splitting, would be nearly twice as great at the sun's poles as at the solar equator.

Hale found in 1908 and 1911 suggestions of Zeeman splitting but not large enough displacements to report with confidence.[17] Construction of a new telescope and a more powerful spectrograph renewed in 1912 the search for the expected Zeeman displacements. Measuring the best of some 288 photographs of solar spectra, Hale's assistants found suggestions of Zeeman splitting. One assistant even found that the displacement was largest at the poles and decreased to zero at the equator, as predicted from theory. But results by different observers on the same plate frequently differed widely, and Hale remained cautious.[18] Attempting to reduce possible systematic errors in measuring, Hale devised another measuring method. The new measures by van Maanen and two others found Zeeman displacements but not definite agreement on an increase in the strength of the magnetic field at higher solar latitudes. With practice, though, van Maanen could measure more consistently, if not always, the expected increase of field strength with latitude.[19]

Aware of the effect of personal bias in scientific observation and measurement, Hale exercised precautions. The plate measurer was to be kept in ignorance of what solar latitude was represented on the plate.[20] This procedure was continued in subsequent measurements, although perhaps with an occasional lapse; a note of qualification sounds in a 1917 report: "the measurer has rarely known in advance the latitude covered by any spectrogram."[21] Van Maanen must somehow have pierced the veil of secrecy, because his overly optimistic measurements of a nonexistent phenomenon continued to find the expected variation of field strength with latitude.[22] An experienced observer, such as van Maanen, could make a good guess about the area of the sun being examined from details in the spectral lines.[23]

Disagreement among different observers kept open the question of a general solar magnetic field. Hale was very anxious to find confirmation of van Maanen's measurements, and in the early 1930s he renewed his efforts.[24] The instrument of renewal was a new measuring technique developed by Hale's English colleague John Evershed.[25]

In 1898 Evershed, then an amateur astronomer, had visited the United States and spent a month with Hale. Evershed was subsequently an assistant and eventually director at the Kodaikanal Observatory in India, where he made observations of the solar spectrum. Upon retirement, Evershed returned to England, where he again met Hale, in 1932. Evershed had just developed a method of measuring spectra that through superpositioning images doubled the observed displacement, thus rendering it considerably more susceptible to detection.[26] Using the new method, Evershed reexamined some of the solar spectra studied by van Maanen in 1914. A few lines were slightly widened, but any Zeeman effect

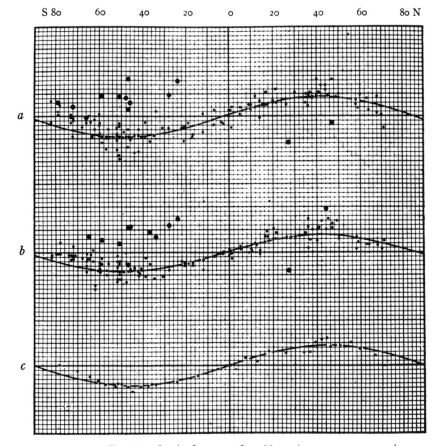

Fig. 7.1. Graph of two sets of van Maanen's measurements, a and b, and a mean curve, c, of the strength of the sun's general magnetic field versus solar latitude. The strength purportedly varies with latitude, as expected. Later measurements, however, found no appreciable general magnetic field, and theory now retrodicts that the polarity in 1912 was the opposite of that reported by van Maanen. Reprinted from George Ellery Hale, "The General Magnetic Field of the Sun," *Astrophysical Journal* 38(1913):27–29.

was too small to separate the lines.[27] After further study Evershed concluded that the observed line widening was not caused by a magnetic field.[28]

Meanwhile, after Evershed's first examination of the spectra but before his final conclusion, Hale was reporting that Evershed had confirmed in most cases van Maanen's results![29] Clearly Hale's anxiety to find the expected result had overcome his scientific judgment.

Hale was not alone in expecting to find evidence of a general solar magnetic field, and Evershed's later report that the line widening was not magnetic in origin was met with strong skepticism.[30] Also contributing to the skepticism was doubt that Doppler effects, the cause of the line widening pointed to by Evershed, could mimic Zeeman displacements, which depend on polarization and are reversed from one hemisphere to the other. There were questions about the magnitude of Zeeman splitting in the solar spectra, but, except in the mind of Evershed, there was little doubt of its reality.[31]

Evershed's report was not the only reason to doubt the existence of a strong general solar magnetic field. On his return to the United States Hale had set five astronomers to measuring recently photographed solar spectra; none of the five could find any displacements of the spectral lines. Perhaps, Hale temporized, the sun's magnetic field varied in strength over time.[32] Another astronomer was then given the task of repeating van Maanen's measures. How strong, one wonders, was the pressure to succeed? Hale's subordinate duly reported verification of the existence and approximate strength of the sun's general magnetic field previously found by Hale and van Maanen.[33] In the published paper no precise number was given for the remeasured field strength, and in fact the value was much less than that asserted earlier by van Maanen. Closer agreement would have been achieved had Hale's often-quoted 1913 estimate of about 50 gauss for the field strength been reduced to 10 gauss, the value that an average of all Hale's spectral lines would have yielded.[34]

Hope of decisive confirmation of van Maanen's solar measurements flickered a final time in the mid-1940s, fueled by emerging reports of wartime research in Germany.[35] G. Thiessen at the

Fraunhofer Institute had developed yet another technique for measuring Zeeman displacements, and he reported a field strength of 53 plus or minus 12 gauss, in excellent agreement with Hale's reported value of about 50 gauss.[36] Several years later, however, having increased the accuracy of his measurements, Thiessen found almost no general magnetic field.[37] Rather than explain the discrepancy between early and later measurements, Thiessen avoided mentioning the initial result. H. von Klüber, working at the Cambridge Observatories, also found no measurable general solar magnetic field and was unable to explain van Maanen's measurements.[38] Astronomers at Mount Wilson failed, too, to confirm decisively the existence of a strong general solar magnetic field. Their initial, loyal conclusion was that the strength of the field varied, from strong when measured by Hale and van Maanen to weak in the 1940s; their more considered opinion was that there is not a strong general solar magnetic field.[39]

The new measurements, which found no detectable general solar magnetic field, raised a serious question about Hale's and van Maanen's earlier results. The pitting of one measurement against another, however, would have done little to settle the matter. Astronomers generally followed the more practical if philosophically unsatisfactory course of favoring one measurement, the newer, and ignoring contrary older measurements.

It was theory rather than observation that finally settled the issue; it was theory that identified the incorrect measurement and prepared the way for its rejection. In addition to the variation in strength with latitude, the general solar magnetic field is positive in one hemisphere and negative in the other, and this polarity switches back and forth over time. Theory predicts, or retrodicts, that the polarity for the period 1912–1916 was the reverse of that reported by van Maanen.[40] Hence van Maanen's measurements are a discrepancy and must be incorrect. To eliminate the discrepancy Jan Stenflo, while a visiting scientist at Mount Wilson, remeasured the solar plates previously measured by van Maanen. Stenflo found no significant magnetic field at any latitude. Current theory was preserved; the sun's general magnetic field had not been essentially different in 1914 from what it was now.[41]

Stenflo also concluded that "systematic errors due to some

personal bias" can easily appear.[42] The forthright attribution of the measurement of a nonexistent phenomenon to personal bias is unusual in scientific literature. It may be significant that Stenflo was a visiting scientist, not a member of the Mount Wilson staff, and also a foreigner.

Scientists may reasonably hesitate to raise the issue of personal bias because of the criticism implicit in the charge. The personal and scientific integrity of the unconscious victim of bias are impugned. Yet the integrity of science itself is impugned by a conspiracy of silence. The dilemma could be avoided were personal bias recognized as an integral part both of human nature and of science produced by humans, were manifestations of preconception accepted if not necessarily condoned.

In the solar case, astronomers had wondered how van Maanen could have found a magnetic field varying in strength with latitude when such a field does not exist, if Hale had really succeeded in keeping secret the latitude. Seth Nicholson, an astronomer at Mount Wilson who had handed some of the plates to van Maanen for measurement and had kept the record of which hemisphere the plate was from, told a colleague in 1951 that a trained observer could make a good guess as to which area of the sun the plate was of by examining "shading" along the spectral line.[43] The immediate response of Henry Norris Russell at Princeton, upon hearing of Nicholson's comment, was to reject any imputation of dishonesty on van Maanen's part. Russell could vouch for van Maanen's complete integrity and was sorry that the controversy over internal motions in spiral nebulae was being coupled with the solar observations.[44] Russell's reflex defense of van Maanen occurred before the 1970 remeasurement of the solar plates.

Van Maanen's asserted measurements of a solar magnetic field and the subsequent reported verifications, sometimes retracted later, do not show science at its best. More skepticism of the initial results, less haste by others to find what was now "known," and a more full reporting of actual measurements rather than vague statements of results "similar" to those found earlier would have been more in keeping with the set of values to which the scientific community aspires.

Nor was it completely proper scientific procedure to ignore

earlier, contradictory results once the absence of a measurable magnetic field became accepted orthodoxy. No doubt it was a practical response in this case, and a response sanctified by subsequent "knowledge," but how can scientists be certain which set of measures is better until the discrepancy is explained? Perhaps efforts to explain the discrepancy were discouraged not only by confidence in the newer measures but also by the fact that one of the potential answers, the effect of personal bias, was not pleasant to contemplate.

While the spectre of a large solar magnetic field ultimately was dissolved, it was as much the discrepancy that rose between observation and theory with regard to the polarity of the field as the scientific community's insistence upon the repeatability of observation and measurement that achieved the result. "All's well that ends well" is a catchy title for a play, but it is scarcely an acceptable justification for less than ideal scientific practice.

8

The Purported Rotation
of Spiral Nebulae

YET ANOTHER twentieth-century failure of objectivity occurred with the purported measurement of the rotation of spiral nebulae from comparisons of photographic plates by Adriaan van Maanen at the Mount Wilson Observatory and by other astronomers at other observatories. The measurement also was central to the great debate over whether spiral nebulae are extragalactic systems.

Philosophical visions arose of island universes, groupings of stars scattered through space.[1] Eighteenth-century belief in the orderliness of the universe made determination of that order an important theological, philosophical, and scientific endeavor.[2] Not until William Herschel's observations late in the century, though, were speculations subject to observation. First, Herschel found many nebulae, cloudy patches in the heavens, resolvable into individual stars with his great telescopes; hence they all were island universes.[3] Then he encountered a nebula he could not resolve into stars; hence no nebula was an island universe.[4] The pendulum of opinion swung again when the Earl of Rosse resolved without exception over fifty of the brightest nebulae, and swung back when William Huggins found gaseous rather than stellar spectra for about a third of some seventy nebulae.[5] Arguing further against the stellar character of nebulae were observations of novae, stars that flare up in brightness; if the nebulae were composed of stars, the novae would have to be many times greater in brightness than the average star.[6] Astronomers now recognize that novae are different from other stars and that there are different types of nebulae, some composed of stars and some composed of gas. The philosophical undesirability of a multitude of types, though, historically furnished strong argument against the island universe theory.

Early in the twentieth century V. M. Slipher's discovery of the extraordinarily high velocities of several spiral nebulae favored an extragalactic nature, since all galactic objects then known had relatively low velocities.[7] The import of his discovery was less clear, though, when a few stars in our galaxy with large velocities were found. And the extragalactic theory was directly challenged by van Maanen's purported measurement of the rotation of spiral nebulae.[8] As Harlow Shapley at Mount Wilson and later the Harvard Observatory argued, if spiral nebulae were extragalactic systems, they presumably would be comparable in size to our galaxy; but if the spiral nebula M101 were even one-fifth as large as the estimated size of our galaxy and rotating with a period of 85,000 years, as reported by van Maanen, the outer edge of the nebula would have to be traveling faster than the speed of light.[9] By a reductio ad absurdum argument, the purported rotation contradicted belief in spiral nebulae as extragalactic systems.

Van Maanen's measurements were not easily dismissed. They were made at Mount Wilson, the world's premier observatory, by a respected astronomer carefully following scientific procedures. Van Maanen's work commanded respect, even acceptance were it not for contradictory evidence of similar heritage. Equally sanctified were the observations of Edwin Hubble, another member of the Mount Wilson Observatory staff. His detection in 1924 of Cepheid variable stars in several spiral nebulae placed the nebulae beyond the boundary of our galaxy as decisively as van Maanen's data confined the nebulae within our galaxy. The disparate findings of Hubble and van Maanen created a dramatic conflict and a difficulty for the Mount Wilson Observatory, the common institutional home for major opponents on a major scientific problem.

A factor in the acceptance of van Maanen's work and probably motivating his research program was the belief that spiral nebulae rotate, though not necessarily rapidly enough to be detected by the comparison of photographs taken a few years apart. The belief was rooted in philosophical considerations and had bloomed into spectroscopic detections of rotation, the first by Slipher at the Lowell Observatory. His report set off a scramble to find rotation for all types of nebulae. When van Maanen reported his photographically determined rotation of a spiral nebula in 1916, he cited Slipher's spectroscopic measurements and papers by other astronomers as well.[10] Presumably to avoid suspicion that preconception might have affected his results, van Maanen later described his discovery as unexpected.[11] The citations in his 1916 paper reveal that it was not. Nor was van Maanen alone; prior to 1916 Carl Lampland at the Lowell Observatory and H. D. Curtis at the Lick Observatory had also attempted to measure the rotation of spiral nebulae from comparisons of photographs.[12] The independent and simultaneous efforts of van Maanen, Lampland, and Curtis indicate that the time was ripe early in the twentieth century for the attempted photographic detection of the rotation of spiral nebulae.[13]

Theoretical considerations possibly also guided van Maanen's work. Thomas Chamberlin, an American geologist, saw photographs of spiral nebulae and thought that their form indicated motion consistent with his planetary hypothesis.[14] Comparison of photographs for evidence of rotation seems not to have occurred to Chamberlin, but he did spend the summer of 1915 at the Mount Wilson Observatory helping prepare an interpretation of the dynamics of nebulae to guide research. Perhaps his theory motivated research at Mount Wilson, though a direct link between Chamberlin's theory and the initiation of van Maanen's measurements is yet to be forged. George Ritchey, a colleague, asked van Maanen to check two plates of the spiral nebula M101 for signs of motion. Ritchey, more a technician than an astronomer, did not publish a discussion of the resulting measurements, and his motives are not known. Van Maanen might have intended to check Chamberlin's theory; he discussed his subsequent results in the context of the theory. In a draft of his paper, though not in the published version, van Maanen wrote that Chamberlin in a personal letter had stated that he saw nothing in van Maanen's measurements inconsistent with the planetary theory.[15] There is a link between van Maanen and Chamberlin, certainly after the supposed detection of motion or rotation in the spiral nebula M101 and probably before. Observational determinations, instrumental capabilities, and theoretical speculations all suggested by 1915 that attempts be made to detect photographically the rotation of spiral nebulae.

Van Maanen had joined the Mount Wilson Observatory in 1912 to measure proper motions and parallaxes of stars on photographic plates. He superimposed with a stereocomparator the images of two photographs of the sky taken at different times. The images of stars with large proper motions, changing positions in the sky from one year to the next, would not coincide. The number of turns of a knob to move a thread from one image to the other translated into a measure of proper motion. Van Maanen obtained proper motions for over 500 stars and for various other celestial objects, including spiral nebulae.[16]

When a few preliminary measurements of nebulous points that van Maanen felt he could identify from one plate to another suggested internal motion in M101, he measured more points. He also obtained additional plates of the nebula taken at the Lick Observatory. To eliminate any relative displacement of points due to the measuring instrument itself, van Maanen interchanged plates from one side of the stereocomparator to the other.[17]

Because the eye compensates for atmospheric fluctuations and for the motion of the object relative to the observer, photographs of celestial details can be less satisfactory than visual observations. Photographs, however, are necessary to bring out details too faint for the eye to register. Ritchey improved photographic techniques to capture some of the advantages of visual observations. Fast-acting camera shutters allowed him to build up a long exposure from many short exposures taken at moments of little or no atmospheric fluctuation.[18] When van Maanen's measurements were later discredited, Ritchey's photographic techniques were suspected of having introduced a systematic error into the data.[19] Nothing to do with Ritchey's plates, however, can account for van Maanen finding similar motions from his examination of the Lick Observatory plates.

Van Maanen chose comparison stars on the plates against which to measure any possible displacement of points within the spiral nebula. The comparison stars were close to M101 on the plate for convenience in measuring but far enough distant to ensure that they were not members of the nebula sharing its motion. The comparison stars also were chosen to be distributed as uniformly as possible around the nebula. And they were chosen to have approximately the same magnitude or brightness in order to reduce the possibility of a magnitude error.[20] A star's image on a photographic plate builds up asymmetrically and varies with the magnitude, as van Maanen had noticed earlier.[21] Despite his precautions, indications of a magnitude error were found in a later examination of van Maanen's work.[22] A random magnitude error alone, however, cannot explain how van Maanen consistently found rotations of approximately the same amount for seven spiral nebulae always in the same direction, with the spiral arms always unwinding.[23]

Van Maanen took care in choosing the comparison stars. He also took care in selecting points in the nebula to measure for changes in position relative to the comparison stars. He rejected nebular points with no surrounding nebulosity because they might be stars in the line of sight between the earth and the nebula.

There remained the problem of measuring the positions. The nebular points were not perfectly symmetrical, and it was difficult to bisect their images and measure the position of the center of each image.[24] Measurement of a pair of plates took several hours, and van Maanen was careful to ascertain that the temperature in the room varied so slightly that any error caused by thermal expansion of the plates or the measuring machine would be negligible. Several of the measurements also were duplicated, to see if any shift in position could be detected. Van Maanen found no evidence of errors. Furthermore, the pairs of plates were placed in the measuring machine in each of four possible positions and the resulting measures combined.

From his measurements of the nebular points van Maanen concluded that the whole nebula was moving and that there were significant internal motions. One component of motion measured from the Lick plates only was three and a half times greater than the same motion measured from the Mount Wilson plates only, while another component of motion was half as much. Van Maanen felt that the agreement was as good as could be expected. Agreement was less good in another test. Internal motions in the nebula were anticipated to be nearly symmetrical with the center of the nebula, but van Maanen found them to be asymmetrical. He rationalized that the points were not uniformly distributed. Agreement was not significantly better, however, when he considered only the points within 5 seconds of arc of the center of the nebula. There were inconsistencies, but this was the best van Maanen could do with the material at hand.

Van Maanen published a photograph of the nebula with an arrow drawn from each nebular point proportional in length to the amount of motion and in the direction of the motion measured. If taken at face value, the indicated motions seemed to van Maanen to represent either a rotation of the points about the center of the nebula or, possibly, motion of the points along the arms of the spiral. A final opinion would be determined by further work.

Van Maanen devised tests to check further the accuracy of his results. Given the uncertainty of some of the measures and internal disagreement among measures of particular points on different sets of plates, confidence in the results was weak. One confidence-boosting test excluded points for which the measures on different pairs of plates differed the most. The apparent rotation easily survived this test. The pattern seemed significant, especially since the comparison stars did not show a similar preponderance of right-handed and outward motion. In a second test the nebula was divided into concentric rings. If the nebular points were moving about the center of the nebula in elliptical orbits, their velocities would decrease with increasing distance from the center. Van Maanen found that the mean rotational components did decrease from the center outwards for all the points and for the more select group, though the decrease was small and hardly trustworthy.

Van Maanen concluded his paper with a list of precautions important when using photographs to study internal motions. Several photographs should be used. Long and short exposures were desirable, short for the bright central regions of a nebula and long for the fainter, outer regions. Pairs of plates should be measured in a differential manner, for which the stereocomparator was ideal. And, to test for systematic personal or instrumental errors, plates should be measured by a second person with a second measuring machine.

Van Maanen did have a colleague at Mount Wilson, Seth B. Nicholson, remeasure parts of the plates with two different machines. Nicholson found the east side of the nebula moving northward relative to the west side by an amount of .083 seconds of arc per year, in satisfactory agreement with van Maanen's value of .063. Next, Nicholson measured 53 points and 32 comparison stars. His measures agreed with van Maanen's about three times out of four on the direction of the motion, and the amount of translational motion was in very satisfactory agreement, + .003 compared to van Maanen's + .002 and − .013 compared to van Maanen's − .016. Nicholson's rotational component was considerably smaller than that measured by van Maanen, + .009 compared to + .022, but this disagreement was not greater than the uncertainty of the data.

Van Maanen's internal motions or rotation, so controversial a decade later, initially created little excitement. The measurement was seen as little more than the careful verification of a phenomenon predicted by theory and independently confirmed by other types of observations. Scientists knew that spiral nebulae rotated. In the 1916 director's report for the Mount Wilson Observatory, Hale placed no special emphasis on the measurement of internal motions in spiral nebulae.[25] He mentioned van Maanen's measurements of the parallax of stars before the work on spiral nebulae, an order probably indicative of the perceived values of the two tasks. The next three Mount Wilson reports noted that van Maanen was continuing his stellar parallax work and obtaining also the parallaxes of some nebulae but did not mention van Maanen's ongoing study of internal motions in spiral nebulae.[26] Not until 1920 did there appear a brief passage acknowledging van Maanen's results for the spiral nebula M33, results published the following year.[27]

Prior to van Maanen's initial photographic study, spectroscopic evidence placed the rotation of spiral nebulae nearly beyond dispute. Shortly afterward, new evidence showed that without doubt a nebula rotated. Francis Pease at Mount Wilson, who had earlier failed in an attempt to verify Slipher's reported spectroscopic rotation of the Andromeda Nebula, obtained during August, September, and October 1917 a 79-hour exposure of the Andromeda Nebula and now found rotation.[28]

More direct confirmation of van Maanen's photographic work

Fig. 8.1. A photograph of the spiral nebula Messier 101, taken at the Mount Wilson Observatory by G. W. Ritchey in 1910.

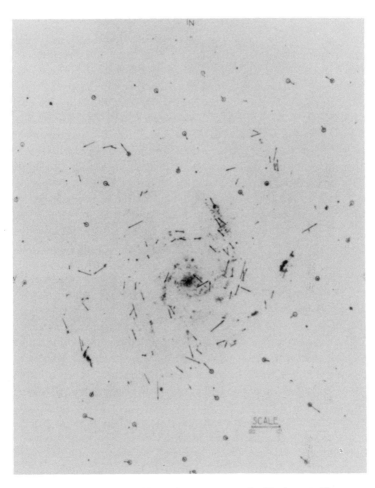

Fig. 8.2. Van Mannen's measurements for Messier 101. The arrows indicate the directions and magnitudes of the purported motions, now known to be nonexistent. The purported motions matched the observer's expectations. Source: Adriaan van Maanen, "Preliminary Evidence of Internal Motions in the Spiral Nebula Messier 101," *Astrophysical Journal* 44(1916):210–28.

also was seemingly forthcoming. Nearly simultaneously with van Maanen's measurement, Lampland at the Lowell Observatory reported the photographic detection of internal motions in the spiral nebula M81.[29] After publication of van Maanen's paper on M101 in 1916, there appeared reports of internal motions in the spiral nebula M51. W. J. A. Schouten, a Dutch astronomer, determined the annual motions of 11 knots in the nebula. Often it was difficult to see which knots on one plate corresponded to knots on the other plate. But the time difference, from 1891 to 1915, was great and the measuring machine very stable. It seemed likely to Schouten that the displacements he found were real. Generally they were larger than his estimate of the probable error. Of the 11 knots, 8 had an inward motion and 3 an outward motion. The mean rotational component of all 11 knots was 0.000 seconds of arc. If, however, 2 knots with anomalously high positive rotational components were excluded, a mean rotation of $-.0073$ was obtained.[30] This was much smaller than van Maanen's value of .022 for M101, but was in good agreement, in Schouten's opinion, with observations of M51 by another astronomer. The Russian astronomer Sergey Konstantinovich Kostinsky had become interested in M51 because it was the best illustration of Chamberlin's planetary hypothesis and it had a long history of observation. Between 1896 and 1916 Kostinsky took 12 plates of the nebula, with exposures ranging from 47 minutes to 3 hours. Comparing stereoscopically an 1896 and a 1916 plate, he noticed almost indisputable displacements of some characteristic knots.[31] Although Kostinsky hoped that a detailed working up of his material would furnish proof of his preliminary results, he seems not to have published further work on motions in M51.

More impressive than the tentative reports of Schouten and Kostinsky were further reports from van Maanen. Ritchey's plates of M101 covered only a five-year interval, from 1910 to 1915. Van Maanen tried in 1920 to obtain another plate, to double the interval. Bad weather during the spring hid M101 from view, but in late summer John Duncan at Mount Wilson obtained a good plate of another spiral nebula, M33. With numerous starlike condensations, M33 was more suitable for measurements of internal motions than was M101, and van Maanen had a good plate of M33 taken by

Ritchey in 1910, as well as two plates of short exposure taken in 1915 and 1920.

Van Maanen measured the plates in a Zeiss stereocomparator. With this instrument he could bisect in quick succession the corresponding nebular points on two plates and thus come closer to choosing the same center point of what were asymmetrical images. Use of the stereocomparator was not without controversy, though. Van Maanen's "magnificent" parallax measurements with the stereocomparator had suggested to his Dutch compatriots that they should try the instrument. One astronomer at Groningen used a stereocomparator with unsatisfactory results. Then another tried, measuring repeatedly distances between different images of the same star on one plate and comparing them to the distances between images of the star on different plates. He should have obtained the same value in each case, but the stereocomparator gave values more than twice as high. The stereocomparator was too sensitive to temperature changes, and all precautions to protect it from changes were in vain. Expansion of different parts of the machine due to temperature changes as small as a few tenths of a degree was intolerable. The Groningen stereocomparator was unsuitable for precise differential measurements. The Dutch astronomers were not prepared to comment on the Mount Wilson stereocomparator. "Still," they thought, "it would seem very hazardous to use any such instrument for this kind of measure unless it has been thoroughly tested."[32]

Van Maanen realized that the stereocomparator was not originally designed for precise measurements, and he did test it. At the Yerkes Observatory in 1911 he had compared results from an ordinary measuring machine and a stereocomparator and had been satisfied with the consistency of the measurements.[33] At Mount Wilson when he began his measurements of stellar parallaxes van Maanen had determined that the stereocomparator was usable if precautions were taken to prevent large temperature changes.[34] He implied in his 1921 paper on M33 that restricting temperature changes to a few tenths of a degree was sufficient precaution. Also, he measured points in one order and then reversed the order in an attempt to eliminate any temperature effect.[35] Still, van Maanen kept after Hale to have an improved stereocomparator constructed in the Mount Wilson shop with the principal sources of trouble in previous instruments removed.[36]

Not waiting for a new stereocomparator, van Maanen went ahead with measurements of M33. The motions were about 80 percent right-handed and 20 percent left-handed, about 90 percent outward and 10 percent inward. More measurements would be required before van Maanen could determine if the motions varied with distance from the center, as true rotation would. He concluded that he had found motions in M33 analogous to those occurring in M101, M81, and M51.[37]

M51 was van Maanen's next subject. He had plates taken by Ritchey in 1910 and a 1921 plate by Duncan. Reasonably consistent results were forthcoming from three different determinations of the mean motion of the nebula as a whole relative to the comparison stars. The rotational period was approximately 45,000 years. The motion outward was in good agreement with that found for M101 and M33. It was so large that van Maanen could not believe that he was observing only rotation; motion outward along the arms of the spiral seemed likely, though further measurements would be necessary before definite conclusions could be drawn.[38]

Next was M81. Van Maanen had 1910 and 1916 plates by Ritchey and a 1921 plate by Duncan. The comparison stars were somewhat brighter than the nebular points, raising the possibility of a magnitude error. The magnitudes of the nebular points, however, had a nearly symmetrical distribution about the center of the nebula; thus the internal motion did not seem affected by a magnitude error. Van Maanen's measuring techniques were the same as in his previous papers and the results were similar. He found a rotational period for M81 of about 58,000 years, compared to 45,000 years for M51, 85,000 years for M101, and 160,000 years for M33. All four nebulae showed large outward radial components, and van Maanen still felt that the data better agreed with a hypothesis of outward motion along the arms of the spirals than with rotation of the nebulae.[39]

The cumulative impact of van Maanen's work was impressive. In the course of the year 1921 Harlow Shapley, who had moved from Mount Wilson to the Harvard College Observatory, found van Maanen's measures of M51 "more convincing than M101 or even

M33"; next congratulated van Maanen on his results for M81; then asked which spiral nebula would be his next victim; and finally reported: "I think that your nebular motions are taken seriously now." At a meeting at which Shapley had discussed the motions, only one astronomer "dared raise his head . . . and later he came around."[40]

The number of spiral nebulae in which internal motions supposedly had been detected increased to seven by 1923, and van Maanen had nearly exhausted his opportunities.[41] Only two or three other spiral nebulae had been photographed by Ritchey between 1910 and 1912, and their images were not sharp enough for comparison with new plates. It was time to summarize and discuss the results of the preceding seven years, and to review thoroughly possible sources of error.

It was unlikely that the purported motions were the result of an instrumental error. The same telescope had produced the old and new Mount Wilson plates, and comparing the plates should have canceled out any instrumental effect. Plates taken at Lick also showed motions. Furthermore, Lampland, Schouten, and Kostinsky using different telescopes all found motions, so they said. Nor was the measuring instrument at fault. Van Maanen had used three different measuring machines. Also, any problem would have affected the comparison stars as well as the nebular points, and the effect would have been canceled out in the reductions of the measures. Furthermore, it was highly unlikely that either telescope or measuring machine could have produced spirals all unwinding regardless of how they were aligned on the photographic plates.[42] Also arguing strongly against an instrumental origin of the measured motions was van Maanen's result when, two years later, he examined plates of the star cluster M13. Had the motions been instrumentally induced, they should have been found in M13 as well as in spiral nebulae, but they were not.[43]

Another potential source of error was the human measuring the plates. Van Maanen made most of the measures himself, but he did have Nicholson make a few check measures on M101 with two different measuring instruments. According to van Maanen, Nicholson had made enough measures "to avoid any doubt as to the result."[44] In a private report to Hale, though, recently discovered by the historians of astronomy Richard Berendzen and Richard Hart, van Maanen conceded that Nicholson had had some difficulty in making the measurements due to eye fatigue.[45] Nicholson's results were within the uncertainties of the material and did not definitely indicate rotation. Despite van Maanen's claim, Nicholson's results did not "avoid any doubt as to the results," according to Edwin Hubble, also at Mount Wilson.[46] Hubble's critique was stinging but suppressed, never to be published.

Van Maanen also claimed verification of his measures by Knut Lundmark.[47] To the extent that both van Maanen's and Lundmark's values were very small, there was some mutual corroboration. But the agreement was over the proper motion of the spiral nebula M33 as a whole, not over internal motions. Lundmark, in fact, distrusted van Maanen's measurements and did not use them to calculate distances to spiral nebulae.[48] Shapley noticed the omission and wrote to Lundmark to ask why he had ignored van Maanen's measures of rotation in spiral nebulae.[49] Lundmark replied that he had mentioned van Maanen's work in other papers, but in the 1921 paper it had been his intention to show that it was hard to give up the hypothesis that the spiral nebula M33 is composed of stars and that the distance is rather large (larger than permitted by van Maanen's measures). While Lundmark did not yet have a definite opinion about van Maanen's internal motions, it was hard not to think that some systematic errors had entered into the measures.[50] He would, though, mention van Maanen's results in the next issue of the journal.[51]

Reporting to van Maanen on the exchange of letters with Lundmark, Shapley concluded that Lundmark had no reason to suspect any systematic error in van Maanen's results but was still lured by his own hypothesis of great distances for the spiral nebulae. Also, Shapley feared that he had hurt Lundmark's feelings.[52] Van Maanen talked with Lundmark and found him "frightfully set in his ways" and unwilling to give up his island universes as yet. Van Maanen did not think Lundmark an easy person to get along with.[53]

Stimulated, perhaps, by the correspondence with Shapley and the discussion with van Maanen, Lundmark published in 1922 a paper more directly critical of van Maanen's work. In 1921 Lundmark had noted that his estimate of the distance of M33 from spectroscopic data was "not in harmony" with the distance estimated by van Maanen from internal motions. Now in 1922 Lundmark continued with a general critique of van Maanen's work, emphasizing inconsistencies. Most serious were differences internal to van Maanen's own work. Lundmark showed that van Maanen's translational motions suggested distances for the spirals six to ten times greater than the distances derived from van Maanen's internal motions. Distance estimates from other data were even larger.[54]

Reacting to Lundmark's critique, Shapley confided to van Maanen that he, Shapley, could talk for an hour or so concerning the frailties of Lundmark's discussion, and he wanted to know how seriously people were taking it. He did not wish to enter into any controversy in these matters.[55] Soon Shapley wrote to Lundmark and learned that Lundmark would stop at Harvard on his way back to Sweden. Shapley looked forward to talking over the matter of Lundmark's statements on the scale of the galaxy, many of which made fine points but others of which threw doubt on Shapley's conclusions regarding the scale of the galaxy without any justification. After "a passing reference or two" to Lundmark's statements, references which took up more than a page in his letter, Shapley concluded:

I have taken more space in writing you of these matters than I had intended, but I feel sure you would prefer to have me write to you directly in a letter. Please understand that I am solely interested in getting at the truth; I certainly do not want to incite antagonism, nor do I object to just criticism of my work. I am interested to have you see that there is a possibility that your treatment might be considered unfair by those who have read the whole literature of the subject and understand the weight of the various arguments. For instance, I heard [Henry Norris] Russell's statement concerning your paper, and I am sure you would not like to hear it yourself. And as you know, he admires your work in general.

With both of us working constructively and observationally on this problem I am sure we will make headway and attain approximate truth. But I think there will be little gain if either of us, or if other students of the subject, strive to pick to pieces small and irrelevant points. It is not of much importance to find small errors here and there if they are not of significance in the larger problem. Think how many flaws or hasty conclusions you or I might find in your big paper on the distances of globular clusters. I have found several small numerical errors, or errors of judgment, in my cluster papers.[56]

Soon after this letter to Lundmark, Shapley wrote to van Maanen to find out if he had again hurt Lundmark's feelings.[57] He had.[58]

Shapley had also, intentionally or not, put pressure on Lundmark to be silent about things in the work of other scientists that seemed not quite right, lest Lundmark invite criticism of his own vulnerable work. The maxims about throwing the first stone and forbidden activities for people living in glass houses, however, were never intended to apply to scientists in their vigorous pursuit of the truth. Shapley's evaluation of certain errors as small, of no significance in the larger problem, rested on the unwarranted and unscientific presumption that he was already in possession of the ultimate truth, that he could judge which observations were significant and which were not.

Nine months later Lundmark replied to Shapley's letter. Although tempted to exchange letters, Lundmark had finally concluded that it would be better and more profitable to talk instead. Now Lundmark was on his way home to Sweden and would stop at Harvard.[59] Apparently the meeting went well. Shapley wrote to van Maanen that Lundmark had agreed to accept van Maanen's measures at present and had said that he did not believe "very heavily" in the island universe theory.[60]

If not a heavy believer, Lundmark nonetheless remained a believer, and it was improper of van Maanen to claim Lundmark's results as corroboration of his own work. Furthermore, van Maanen's 1923 paper in which he claimed corroboration was dated two months earlier than the conciliatory meeting between Lundmark and Shapley.[61] The inconsistencies that Lundmark had pointed to were real; his work could not legitimately be claimed as confirmation of van Maanen's results. Hubble, critical of van Maanen's claim of corroboration from Nicholson's measurements, was even more critical of van Maanen's claim of support from Lundmark's work. The numerical values of van Maanen and Lundmark were, according to Hubble, irreconcilable.[62] This remark, too, was not published.

In addition to the claimed corroboration of Nicholson and Lundmark, van Maanen also cited the measures of M51 by Lampland, Schouten, and Kostinsky. Lundmark had noted that these measures were not entirely consistent.[63] Hubble's unpublished criticism was more sharply drawn. According to Hubble, Lampland's results were "entirely vitiated" by his enormous proper motion for the central nucleus of M51.[64] As for Kostinsky and Schouten, their measures were without any significant weight, but in so far as the measures suggested rotation, it was in a counterclockwise direction, *opposite* to van Maanen's clockwise rotation. Van Maanen's reference to Kostinsky's and Schouten's reports of "similar motions" indicated to Hubble an "uncritical imagination."[65] Privately, Hubble pointed out to van Maanen the "utter impropriety" of his claim. According to Hubble, again in a manuscript never published, van Maanen admitted to Hubble that Schouten's motions were in the opposite direction and that Kostinsky's motions did not support him.[66] Made in private, Hubble's criticism and van Maanen's possible partial retraction had little effect on scientific opinion.

Van Maanen's measured direction of rotation also was contradicted by spectroscopic determinations, a fact not acknowledged by van Maanen in print, though he was aware of the problem. Upon finishing his measurement of M81, van Maanen had written to Shapley that M81 was beautiful, with motions like the other spirals, but if Slipher were correct in his direction of rotation, then his own measures must be wrong. Maybe, though, a reversion of the direction of motion somewhere near the center of the nebula could explain the apparently incompatible results.[67] Shapley seems either to have ignored the problem or to have accepted the improbable suggestion of a reversion of motion, a suggestion van Maanen himself punctuated with three exclamation marks. Three months later, following a visit with Mount Wilson astronomers, Shapley wrote to van Maanen that he thought the internal motions were taken seriously now.[68]

A few other astronomers also noticed the discrepancy in the reported directions of rotation, as the historians Berendzen and Hart have shown, but did not publish the fact.[69] The scientific community holds as an ideal a vigorous and critical examination of knowledge and relies on scrutiny after the fact, along with proper procedure in determining facts, to weed out errors. The few scientists who noticed the obvious discrepancy between the photographic and spectroscopic rotations of spiral nebulae and the timidity with which they pursued the issue suggests, however, that the ideal of vigorous and critical examination is not always completely achieved in actual scientific practice.

While there was, over time, an erosion of confidence in van Maanen's work, observations by other astronomers had initially seemed to corroborate his findings. There was also, initially, a satisfying agreement between van Maanen's measures and theories concerning the spiral nebulae. Van Maanen had raised the issue of Chamberlin's planetary hypothesis. James Jeans, an English astronomer, developed that hypothesis further, and van Maanen's measures were even more compatible with Jeans's model.[70] In a 1917 note for the Royal Astronomical Society on the recent progress of astronomy, Arthur Eddington remarked upon the spectroscopic observations of rotation of nebulae, upon van Maanen's very interesting results, upon the apparently confirming work of Kostinsky, and upon the fit of observations with Jeans's dynamical theory.[71] Jeans, himself, at a 1921 Royal Astronomical Society meeting showed slides of van Maanen's purported internal motions in spiral nebulae, called upon the viewers to "admire the consistency of measurements," and reported that the measures confirmed the theory underlying his conjecture about the real nature of the condensations in the arms of spiral nebulae.[72]

Van Maanen's observations appeared to fit Jeans's formula for the shape of spiral arms. But when Jeans later analyzed the effect of the force of gravity upon the nebular points, he encountered difficulty. No stronger expression of confidence in van Maanen's results could be imagined than that now put forward by Jeans. Rather than treating the difficulty as a reductio ad absurdum argument and rejecting van Maanen's reported motions, Jeans decided instead that he must look for "some new and so far unknown force for an explanation."[73] Ernest Brown at Yale University took up the problem. He felt that van Maanen's measurements were remarkably accurate and he attempted to explain the observed internal motions by gravitational action alone. Brown succeeded in hypothesizing a particular distribution of matter in spiral nebulae that could in combination with the laws of gravity produce the observed internal motions.[74] It was a possible reconciliation between theory and observation, but a difficult one. Both Jeans and Brown quickly abandoned the effort when other observational evidence from 1924 onward made it apparent that van Maanen's purported measurements were illusory.[75]

However fickle the support from theoreticians, it was not unimpressive at its height. Nor were the first reports from other astronomers of similar measures of internal motions. Nor was the spectroscopic evidence of rotation. Combined with these factors in favor of the asserted internal motions was the force of van Maanen's scientific reputation. He was respected for his careful and accurate work, not only in laudatory obituary notices but while he lived. His efforts to eliminate potential sources of error were evident and impressive.[76] Astronomers at Groningen had been led to experiment with the stereocomparator because of van Maanen's "magnificent" parallax measurements with that instrument.[77] And in 1924 W. M. Smart, an expert at Cambridge on stellar dynamics, expressed his confidence in van Maanen's work in the strongest terms. Smart's own results were identical with van Maanen's and thus "eloquent testimony to the extraordinary precision of the measures." He also wrote, "I do not believe that anyone would be so bold as to question the authenticity of the internal motions— regarded either as rotational or as a stream motion—found by van Maanen; in fact, the more one studies the measures, the greater is the admiration which they evoke."[78]

Intertwined with reputation were personal relationships and beliefs. Shapley later wrote that his faith in the correctness of van Maanen's measures was strengthened by ties of friendship.[79] Correspondence from 1917 suggests that Shapley initially doubted van Maanen's motions. He had written to Russell at Princeton that van Maanen doubted the motions a little, Hale more, and he, Shapley, much. Shapley did, though, express his admiration for van Maanen's ambition, skill, and desire to produce further evidence, if only he had plates to measure.[80] Nor was van Maanen's evidence incompatible with Shapley's beliefs. Shapley was the major opponent of the island universe theory in the early 1920s, and he congratulated van Maanen on his "nebulous results," gloating that: "Between us we have put a crimp in the island universes, it seems—you by bringing the spirals in and I by pushing the galaxy out. We are indeed clever, we are. It is certainly nice of those nebulae to have measurable motions. I wonder if W. S. A. [Walter S. Adams at Mount Wilson] believes it."[81] Shapley's own beliefs may well have conditioned his acceptance of data. And personal relationships were not negligible factors for Shapley. In another instance, in which he had refused at first to accept that nebulous appearing images were really stellar images, he later explained that he had been impressed by Ritchey's opinion that the images were not stellar.[82]

There were reasons, ranging from personal to scientific, for placing faith in van Maanen's measurements. Scientific correspondence reveals a widespread acceptance in America of van Maanen's results. While Shapley had expressed some doubt in 1917 about the internal motions, Russell had expressed his confidence in the measures, a confidence later repeated.[83] Shapley soon came around and reported that other astronomers were coming around as well, leaving Curtis and Lundmark the only strong defenders of the island universe theory.[84] Van Maanen's results also were accepted in England, perhaps even more so than in the United States. At the December 1924 meeting of the British Astronomical Association, members were told that van Maanen had "measured unmistakeable movements in several of the larger spirals. . . ." In cautioning that "the rapid progress of knowledge, and the changing state of speculative theories of the nature and origin of these objects, perhaps make the compilation of such a paper as the present one rather a risky procedure," the speaker was more prophetic than he realized.[85] In the *Monthly Notices* of the more professionally oriented Royal Astronomical Society, a similarly matter-of-fact report noted that "in the case of the spiral nebulae, the outstanding contribution has been the measurements by van Maanen of the motion of condensations in the spiral arms. . . ."[86]

Van Maanen's purported measurement from comparisons of photographic plates of large internal motions in spiral nebulae was generally accepted by astronomers, by all but a few adherents of the island universe theory, which was contradicted by van Maanen's measures. Corroboration by other astronomers, ready connection to and compatibility with dynamic models of nebulae, and van Maanen's reputation for careful and exact work all made for a favorable reception of his results—and for a difficult situation when incontrovertible evidence incompatible with van Maanen's measurements was produced at the Mount Wilson Observatory.

Hubble's discovery of Cepheid variable stars in spiral nebulae and his use of the Cepheids to determine distances to the nebulae finally settled the long-standing question of the identification of spiral nebulae as island universes. Early in 1924 Hubble wrote to Shapley:

You will be interested to hear that I have found a Cepheid variable in the Andromeda Nebula (M31). I have followed the nebula this season as closely as the weather permitted and in the last five months have netted nine novae and two variables . . . the distance comes out something over 300,000 parsecs [about a million light years]. . . . I have a feeling that more variables will be found by careful examination of long exposures. Altogether the next season should be a merry one and will be met with due form and ceremony.[87]

It was a bitter pill for Shapley to swallow. When Hubble's letter arrived, Shapley said "here is the letter that has destroyed my universe."[88] To Hubble, Shapley replied that his letter "telling of the crop of novae and of the two variable stars in the direction of the Andromeda nebula is the most entertaining piece of literature I have seen for a long time." Shapley was sure that the large number of plates Hubble had obtained furnished adequate assurance that the stars were genuine variables.[89]

By August Hubble had even more variables and was beginning to consider the significance of his data.[90] Shapley also recognized that the straws were now pointing toward the island universe hypothesis and away from van Maanen's measured rotations. He replied to Hubble: "Your very exciting letter received here. . . . What tremendous luck you are having. . . . I do not know whether I am sorry or glad to see this break in the nebula problem. Perhaps both. Sorry because of the significance for the measured angular rotation, and glad to have something definite and interesting come to hand."[91]

Hubble was slow to publish the exciting results. As he explained to another colleague: "The real reason for my reluctance in hurrying to press was, as you may have guessed, the flat contradiction to van Maanen's rotations. The problem of reconciling the two sets of data has a certain fascination, but in spite of this I believe that the measured rotations must be abandoned."[92]

Hubble attempted to deal with the problem of contradictory data in a talk at Mount Wilson. He asserted that only "Mr. van Maanen's measures" contradicted the evidence of the Cepheid variables. There was also evidence from novae, from the observed colors of the brightest stars, and from star counts, all of which helped to build up a consistent picture, "that is save for a disturbing shadow cast by the measures of Mr. van Maanen." Hubble wrote in his notes for the talk, "It is well to emphasize this isolation, for on the validity of rotations depend in any important infringements of that comfortable principle, the uniformity of nature. (The real controversy is between the universality of Cepheid variation and the validity of Mr. van Maanen's motions.)"[93]

Hubble also criticized van Maanen's claims of corroboration. In a brief paper Hubble put his objections bluntly, using language such as "entirely misleading" and "an uncritical imagination."[94] Hubble, however, didn't publish the criticism.

Hubble suspected van Maanen's motions, but how could such an experienced measurer of photographic plates have been so misled as to measure nonexistent rotations? A magnitude error, involving asymmetrical images on photographic plates, was a possibility.[95] For Hubble, "the problem of reconciling the two sets of conflicting data is fascinating but we must be certain on the reliability of the data before we proceed." He was trying to force van Maanen to remeasure the globular cluster M13, for which van Maanen had found no rotation. While Hubble viewed the whole affair as "rather a lark," he wrote to Shapley that "van Maanen takes the idea of a magnitude error in a wildly personal manner."[96] Finally submitting his long-delayed paper on Cepheids and distances to spiral nebulae, Hubble explained: "The delay in publishing the cepheids has been due to the conflict with van Maanen's rotations of spirals. It seems now as if these latter may be explained as a simple magnitude effect, though you can realize how delicate a business such a discussion will be."[97]

Hubble made little progress in the next few years in his search for a magnitude error. No serious opposition to his distance determinations spurred Hubble to complete a refutation of van Maanen's contradictory measurements. Nor had Hubble overcome his reluctance to enter into a public controversy with van Maanen. Also, Hubble soon had more important problems commanding his efforts. Mrs. Hubble later remembered both the period in which van Maanen's measurements could be ignored and how the situation changed:

It was, the contradiction, not very important in the long run. The work [Hubble's distance determinations] had become so obvious and so extensive that van Maanen's measurements could be ignored. . . . Van Maanen had refused to discuss the subject or to re-examine his work. Edwin had decided to ignore the discrepancy for the time and go on with his work. A friend of his told me years afterwards that he [Hubble] said to him "They asked me to give him [van Maanen] time. Well, I gave him time, I gave him ten years." He [Hubble] knew the measurements were wrong, his close colleagues knew. To him, that would be enough. To publish would be to expose van Maanen to public humiliation. The comment he made, his friend said was made without rancour.[98]

That was Hubble's feeling in the late 1920s. Things changed in the early 1930s. According to Mrs. Hubble:

Milton Humason [who worked closely with Hubble] had told me that van Maanen always refused to let anyone see his plates or measurements. I do not remember E talking of it. But his friend and neighbor, Franklin Baldwin told me of two comments E had made to him, partly because he was amused by E's almost astronomical attitude towards human affairs. He [Hubble] said calmly with no appearance of rancour, "They asked me to give him time; well, I gave him time, I gave him ten years." Then he said, "When a speaker at the R.A.S. [Royal Astronomical Society] announced, if it were not for van Maanen's measurements, Hubble's results might be accepted, I decided to make the measurements."[99]

Walter Adams, then the director at Mount Wilson, may have asked Hubble and van Maanen to remeasure the plates to settle the conflict.[100] New measures by van Maanen showed "a decided internal motion in the same direction as was found in his original measures of this nebula [M101]."[101] Hubble, however, comparing Ritchey's plates with new plates taken by Milton Humason, found "no evidence of motion." Nor did Nicholson, who had earlier measured van Maanen's plates, find any motion now. Hubble concluded, in yet another paper never published, that "the new measures indicate no definite evidence of systematic internal motion within the limits set by the probable error." Van Maanen's measures, over twenty times as large, exhibited "a conspicuous correlation between magnitude and rotation. . . ."[102]

Hubble pursued the possibility of a magnitude error in additional manuscripts, also not published. A correlation between rotation and magnitude was strongly suggested but proved elusive to pin down. Reviewing his analyses of M33, M51, and M81, Hubble wrote: "No adequate explanation of van Maanen's personal equation has been found. Study of the plates persistently suggests the unsymmetrically distorted comatic images as possible sources of systematic errors related to magnitude."[103] The correlation between rotation and magnitude was very conspicuous for M81 and M51, was "suggested" in the measures of M33, and was "possibly" present in plates of M101. "If the data for M81 and M51 stood alone," Hubble reasoned, "the correlations would serve as reasonable explanations of the results." But the data did not stand alone. All the rotations were in the same direction, unwinding with respect to the spiral arms. Thus it was difficult to attribute the motions to random errors, since random errors would be expected to give the appearance, randomly, of both winding up and unwinding. Also, van Maanen had obtained negative results for globular clusters. For these reasons Hubble made "no attempt to fully account for the predicted rotations. . . ." He was still attracted, though, to the "bare possibility that magnitude errors, by chance and very improbable coincidence might save the phenomena."[104] There remained the problem that all seven of the spirals reported by van Maanen were purportedly rotating in the same direction with respect to the arms and van Maanen's measures of two globular clusters using the same technique had found no appreciable internal motions. The problem was not too little data but too much: the reported motions of seven spirals all in the same direction was too great to be coincidence.

In the longest of his unpublished manuscripts Hubble launched a final analysis and a last attempt to explain away van Maanen's erroneous results:

Renewed interest in the question is evidenced by occasional references to this outstanding discrepancy in the field of nebular research and by a growing tendency to consider the possibility of constructing theories to reconcile the apparent angular rotations with the currently accepted large distances. For this reason it seems timely to re-examine the data in greater detail and consequently four spirals have been remeasured using the longer intervals now available.

Hubble found no observable internal motions in the four spirals. "Since the presence of concealed errors is established," he concluded, "the significance of the entire set of measures is uncertain until the source of the errors has been identified and their effects removed." Again Hubble focused his attention on the possibility of a magnitude error; again he found that a magnitude error was not a sufficient explanation:

> The actual diagrams [of magnitude versus rotation] appear to favor the latter interpretation [of a magnitude error] in the three cases presented [M33, M51, and M81], but in general they are indeterminant and do not themselves uniquely establish either interpretation. Under such conditions the choice must be determined by the independent evidence which includes the systematic errors established by blinking and the general body of knowledge. Thus the selection of rotation now appears as an arbitrary interpretation not forced on the observer by the data themselves, but read into the data by the observer.

Hubble must have been uncomfortable accusing a fellow scientist of reading an arbitrary interpretation into data, and he attempted to evade this conclusion. Revising his manuscript, Hubble expanded the discussion of the possibility of a magnitude error and more forcefully stated that he had found correlations between rotation and magnitude for M51 and M81. He was forced yet again, though, to conclude: "Nevertheless, in view of the fact that all the rotations are in the same direction with respect to the nebulae and not to the plates, it seems impossible to account for them by magnitude errors alone."[105]

Hubble never published his manuscripts. The Mount Wilson bureaucracy, especially Frederick Seares (the editor for papers written by Mount Wilson astronomers), discouraged Hubble from publishing, as the historian of astronomy Robert Smith has shown.[106] Following a phone conversation with Hale in which Seares had not been able to talk freely because some unnamed person, presumably Hubble, had been in the office with him, Seares set out his feelings in a letter.

> The following sounds academic; nevertheless it states more explicitly than I put it to you yesterday the general principle that has determined my actions in respect to our present troubles.
>
> For two men in the same institution there is opportunity for personal contact and for direct examination of each others results, and for the private adjustment of differences in opinion. The institution itself, it seems to me, is under obligation to see that all adjustment possible is made in advance of publication. If agreement cannot be obtained it may be necessary for the institution to specify how the results shall be presented to the public. In that event, however, there must be opportunity for the expression of individual opinions but any such expression should be concerned only with the scientific aspects of the question at issue.

Hubble evidently didn't agree with Seares, perhaps over the issue of institutional censorship or perhaps over which aspects were scientific. Seares's letter continued:

> The institution has, I think, the right to enforce this procedure; but in certain cases it may be wiser to waive its technical right and say to a dissatisfied individual "Print what you like, but print it elsewhere." Under no circumstances is it obliged to accept for its own publications material prepared in a manner of which it does not approve. Mere acceptance implies a certain sanction; and unless the material presented conforms to standards of dignity as well as of excellence, its acceptance may give the impression of ineffectual direction and cause disruption of morale within the institution itself.

What Seares hoped for was a compromise and a more seemly publication from the institution. Seares's letter continued:

> One detail has cleared itself up only since our talk; and that led us to the suggestion made in my note this morning. To this I have not been able to find any fatal objection. I am very anxious to have your opinion about it after you have had time for consideration—my original proposal of a joint paper I still think is the most desirable solution but I doubt its practicability unless we are willing to put up a fight—that after dragging out for months might end in Washington [the National Academy of Sciences].[107]

The matter was of great concern to Seares, so great that he sent off a second letter to Hale later the same day:

> After further thought about the matter I am disposed to think that

any conference between you and Hubble would fail in the attempt to relieve the difficulty—at least I think we should consult with Adams about it before making the attempt. This feeling is determined largely by the character of Hubble's comments on my notes which I saw yesterday for the first time.

I am beginning to wonder if the solution does not lie in a suggestion to Hubble that he print his investigations elsewhere than in the observatory publications, in order that he may have entire freedom of expression—there are objections, but the difficulties, I think, may be less serious than those sure to be—in any alternative procedure.[108]

Yet another letter followed the next day:

. . . Here is the daily installment—an attempt to put my suggestions in working form:

Returned H's last MS (revised copy in your hands) to him with a notification that the Observatory does not consider it advisable to publish this MS; that it has proposed a means of publication which it considers just and fair to all parties concerned, but, since the proposal is unacceptable to H, the observatory does not insist—that he has entire freedom of decision. Make this a matter of _____, then proceed as though the matter were ended.

If, without notifying the Observatory, H sends his MS elsewhere for publication, make no protest and say nothing.

If he announces his intention of sending his MS elsewhere, say merely, without any protest, that the responsibility for any such decision is his own.

This will give our vital thing that I have held out for from the beginning—everything else seems to me relatively unimportant. But please help me find the notes.[109]

Events were to prove Seares overly pessimistic. Hubble and van Maanen didn't publish a joint paper but they did reach an agreement. Hubble published a brief note, pointing to van Maanen's results as the outstanding discrepancy in the conception of nebulae as extragalactic systems and presenting a table of his own remeasures of the four spirals, M33, M51, M81, and M101. The results established the existence of some systematic errors in the rotations. On the source of the error Hubble was silent.[110] Immediately following Hubble's note in the journal was one by van Maanen, obviously by arrangement. Van Maanen recalled how in his measures of internal motions in spiral nebulae he had been aware of several sources of possible systematic error and had, after publication of his results on M33 in 1923, postponed further investigations until he had available plates from the 100-inch telescope. Now he had such plates, and the internal motions, while still showing the same direction of motion as before, were smaller. The difficulty of avoiding errors, the new smaller results, and Hubble's remeasures (confirmed by Nicholson and by Walter Baade, another astronomer at Mount Wilson) all made it "seem desireable to view the motions with reserve."[111] Van Maanen spoke of future investigations to resolve the issue, but the pair of brief papers in the *Astrophysical Journal* of 1935 marked the official end, if not a resolution, of the one discrepancy in the otherwise consistent picture of the spiral nebulae as galaxies beyond our galaxy.

The real reason for van Maanen's error, that he had read his expectations into his data, was not a satisfactory explanation, even if true. Published, it would have humiliated van Maanen and generated public controversy. Hubble could not describe correctly the error and remain a gentleman scientist; the truth remained unstated, buried in Hubble's unpublished manuscripts.

The scientific import of van Maanen's measurements has passed. Any remaining significance now resides in historical, philosophical, and sociological considerations. The case of the purported internal motions in spiral nebulae throws into doubt the supposed objectivity of science, both the efficacy of the scientific method and the vigor of peer review.

9

Objectivity Questioned

THE CASE of the alleged rotation of spiral nebulae, particularly, and other supporting and amplifying instances in the recent and not so recent history of science suggest that the image of objective science may exceed the substance. The image requires modification, qualification, and perhaps even replacement, no matter how psychologically comforting and culturally necessary has the idol been for its guarantee of human progress.

The ideal of objectivity came in the aftermath of the Great War to characterize science and to distinguish it from other human activity, to raise it above all other human activity. Yet science remains a human activity, subject to inevitable human bias. That bias has proven more clever and more pervasive than all the restraints man has thought up to control it.

Recent advances in historical knowledge of instances in which scientists have found what they expected to find even when it was not there to be found now furnish a basis for a reassessment of the role of observation in science and the concept of objective science. A common pattern or etiology is emerging.

Supposedly objective observation is, of course, subject to a host of difficulties in addition to personal bias. It is all too easy to confuse the effect of personal bias with other factors whose significance lies either in effecting a systematic skewness in the measured data or in creating ambiguities in the interpretation of data. Some of these difficulties need to be mentioned, if only to delineate better the problem of objectivity by stating what it is not.

At one extreme of the spectrum of observational problems is the selection effect. Put simply and in a common astronomical context, some observations are more easily made while others escape the observer's instrument. Bright stars are more readily detected than are faint stars, and observed density distributions must be corrected accordingly. Stated more elegantly by the astronomer and philosopher Sir Arthur Eddington, fishermen using a two-inch net might well conclude from an examination of their catch that no fish was shorter than two inches.[1] There is a subjective element introduced into science by the selection process governing the collection of data, but it is not what led to the spurious results of Adriaan van Maanen and others.

Nor are spurious results to be included within the problem of rival interpretations based on the same data base, or rival emphases based upon different subsets of a common data base. The historian and philosopher of science Thomas Kuhn has shown that the scientific community may at different times recognize different parts of the data base as requiring explanation or as supporting one theory over another.[2] At issue in the case of internal motions in spiral nebulae and other instances of failed objectivity, however, is the reliability of the raw measurements, not their subsequent interpretation. Preconceptions do importantly influence the perception and interpretation of data, but that is a separate issue from the question of the reliability of raw data.

Preconception also helps determine the vigor and thoroughness with which data is gathered. It is difficult to determine when an experiment or observation has been completed, when all the errors have been weeded out and only truth remains. One indication of the status of an experiment or observation is its degree of agreement with the expected result. In the face of disagreement, a scientist's natural response is to check for errors. Should there be early agreement between observation and preconception, a scientist may be lulled into premature complacency.[3] That happened in the measurement of the speed of light. In 1941 it was thought to be 299,776 kilometers per second, with a maximum error of 4 kilometers per second. But new technologies soon increased the accepted speed by 17 kilometers per second, by more than four times the accepted margin of error. The spurious agreement of several sets of pre-1941 measures, one of the most astonishing systematic errors in the history of physics, is now attributed to the awe in which the great Albert Michelson and his measurement of the speed of light were held and to the readiness of several scientists, when their early measurements agreed with Michelson's, to accept that value and end the search for errors. As one reviewer noted:

The tendency of a series of experimental results at a certain epoch, to group themselves around a certain value raises a very interesting psychological question. . . . In any highly precise experimental arrangement there are initially many instrumental difficulties that lead to numerical results far from the accepted value of the quantity being measured. . . . Accordingly, the investigator searches for the source or sources of such error, and continues to search until he gets a result close to the accepted value. *Then he stops!* But it is quite possible that he has still overlooked some source of error that was present also in previous work. In this way one can account for the close agreement of several different results and also for the possibility that all of them are in error by an unexpectedly large amount.[4]

The process has been labeled "intellectual phase locking."[5] No doubt it accounts for some, perhaps many, instances of minor failures of objectivity in science. In the cases of the spurious rotation of spiral nebulae, the general magnetic field of the sun, and the gravitational redshift of Sirius B, however, the initial alleged measurements were too far from reality to be attributed, in retrospect, completely to instrumental errors and absence of due diligence.

Errors do occur in measuring data, and astronomers were among the first scientists to recognize this fact of life. In the eighteenth century, distressed that one of his assistants consistently erred in observing the movement of stars across the sky, the English Astronomer Royal first admonished the poor man, then warned him, and finally sacked the hapless individual. Early in the nineteenth century the German astronomer Friedrich Bessel, more respectful of men, concluded from similar difficulties that there are individual differences in measurements, that there is a personal equation.[6] The problem is not solely that of random errors, extensively studied and known to display a Gaussian or bell-shaped curve centered on the actual value. The problem of a personal equation involves consistent differences between different observers. Skilled observers will note the passage of a star across a wire in the eyepiece of a telescope slightly too late or too early, by an amount that differs for each observer but remains nearly constant for an individual observer observing different stars. It is not an organic problem with eyes or ears, but an acquired habit of observation. And it is easily dealt with. The error is canceled out when an astronomer takes the difference of the times of passage of two stars across the transit wire, because the astronomer makes approximately the same error in each case. Similarly, astronomers cannot be expected to place a cross exactly over the images of stars. One observer will consistently perceive the centers of stellar images slightly to the right of the actual centers, while another observer will pick points slightly left of true, and others will pick slightly high or slightly low. Reversing the photographic plate reverses the error following from the personal equation; taking the mean of two measurements, one too far to the right and the other too far to the left, eliminates the error. The personal equation does not explain van Maanen's problem. He did reverse his plates and take a mean of the different measurements. His spurious measurements demand errors not consistent in relation to the photographic plate but with the expected direction of rotation, not consistent with the measurements but with the hypothesis.

Van Maanen's problems arose from bias in the direction of the hypothesis held by the observer. One result of bias, though not that experienced by van Maanen, is an interpersonal self-fulfilling prophecy. In the Pygmalion experiment, teachers were told that certain students were ready for intellectual growth; eight months later the teachers reported such growth. Presumably the teachers gave the students more or a different quality of attention than they otherwise might have; the bias of the experimenter affected the process under observation. Van Maanen, though, was recording replicable events not subject to his personal influence. Only recently have psychologists moved beyond interpersonal experimenter effects to cases such as van Maanen's. It is too early to say how often measures are biased in the direction of the hypothesis held by the observer. But it is already beyond reasonable dispute that in some cases errors have occurred and that the direction of the errors is correlated positively with the expectations of the observers.[7] The hypothesis, even belief, that scientific knowledge is strictly controlled by observation statements, which in turn are established beyond question by rigorous scientific method, is contradicted by recent psychological studies as well as by the historical record.

One last problem involving scientific observation merits at least brief mention. It is deliberate fraud. Several recent instances of fraud greatly weaken, if not completely destroy, the presumption that scientists necessarily are deterred from fraud by the certainty of eventual detection.[8] There is, however, no evidence of deliberate fraud in the case of the spurious rotation of spiral nebulae, nor in many other cases, and it is unlikely that new evidence will shift all historical cases of failed objectivity in science into the category of deliberate fraud. In at least some instances, and probably in many, personal bias and self-deception have led honest scientists astray. Observations undertaken in good faith have proved an unreliable basis for restricting theory. There is much to be learned about science from a study of recent frauds, not the least of which may be an increased realization of tremendous pressures inherent in a desperate competition for patronage. But the concept of fraud is not sufficient, alone, to explain away all apparent instances of the failure of objectivity in science.

Accompanying good faith is often a high level of scientific skill. Observational errors are not a monopoly of the incompetent. In exoneration, even in praise, of the Mount Wilson Observatory and its embarrassing involvement in several observational fiascos, it might be asserted that the involvement is due more to Mount Wilson's merit than to any deficiency. Much as the skilled surgeon operating on nearly hopeless cases suffers a higher mortality rate than does the general family doctor treating a common cold, so astronomers at the Mount Wilson Observatory observing at the very limits of their magnificent telescopes and scientists everywhere working on the frontiers of science might well be expected to produce more than their fair share of observational errors.

Scientists, of course, are aware of the opportunity for error inherent in observation, probably more aware than are nonpractitioners. The history of science, at least the history of modern science, is full of exemplary instances in which scientists took extraordinary care to eliminate all conceivable opportunity for mistake — and, in too many cases, still made mistakes.

Thomas Harriot early in the seventeenth century working, speculating, or whatever, quite likely beyond his instrumental capability to observe the lunar surface or sunspots with any certainty, had little opportunity to practice care. By the end of the century, however, a Robert Hooke, a John Hevelius, or a John Flamsteed, especially the latter, could make systematic, quantitative measurements with a well-estimated range of error. Furthermore, corrections for effects such as atmospheric refraction were being applied to measurements. Astronomers had set for themselves the goal of precision; their science would become increasingly precise as instruments and observing practices were further improved.

Astronomers in the next century demonstrated an awareness of their own psychological makeup and its consequences for observation. William Herschel noted the phenomenon that a minute object much smaller than what one is used to seeing will at first appear less than it really is. Done in retrospect and used in a futile attempt to explain away an embarrassment, Herschel's sophistry may escape commendation. Nonetheless, it marks an advance in the quest for objectivity.

Herschel also searched, in the case of the changing diameter of Uranus, for errors possibly introduced by his instruments. Again, the search was in retrospect but nonetheless commendable. It was but a small, if important, step from searching for errors after the fact to searching for sources of error before making observations, from explaining away obvious mistakes to preventing mistakes, and also avoiding mistakes both recognizable and not. Herschel took the step himself in his search for a ring around Uranus. Aware that the appearance of a ring might be caused by an imperfection in the telescope mirror, he rotated the mirror. The appearance of a ring was unchanged; hence it was not an artifact of the mirror, Herschel reasoned. It was, however, in some way produced by the telescope, because the appearance of a ring retained its orientation with the telescope tube rather than with the planet over the course of the night's observations, as Herschel intelligently recognized. His efforts to distinguish instrumental artifacts deserve more recognition than they have hitherto received. As good telescopes became more widely available to good astronomers a further check on observation was furnished by the practice of verification by different astronomers using both the same and different instruments. Though verification was too easy to come by for Neptune's supposed ring in the nineteenth century and for other expected phenomena even in the twentieth century, the scientific emphasis upon verification is useful. A shared belief in the existence of the phenomenon sought, however, often precludes the possibility of truly "independent" verification, independent of shared preconception.

The planet Neptune serves also as a reminder that through the history of modern science there runs the important theme of patronage. Observations of Neptune were not uninfluenced by concerns of national and personal advantage, nor were Hooke's seventeenth-century observations, nor are contemporary scientific programs.

Elements of possible error instrumentally induced and human error psychological in origin were well known to scientists by the twentieth century and are readily apparent in the Lowell saga and the Martian canals. The hope then expressed, perhaps more so in the United States than in Europe, that human bias could be avoided if only the observer would place his brains in a blind trust, at least until after making his observations, probably constituted a backward step in the quest for objectivity. Better to accept the inevitable human bias and concentrate on checking its reach than to pursue the chimera of bias-free observation.

Despite critical comments on Lowell's unguarded expressions of preconception, scientists recognized the possibility, if not always the inevitability, that preconception might influence observation. Means of supposedly eliminating such bias and its unwanted influence were practiced at the Mount Wilson Observatory in the search for a general solar magnetic field. When giving van Maanen plates to measure for possible Zeeman splitting and hence a magnetic field, Hale did not tell van Maanen what solar latitude an observation was from. The ensuing determination of a nice correlation of field strength with latitude was most impressive—and now understandable only if van Maanen, consciously or unconsciously, saw through the veil of latitudinal anonymity.

The case of the spurious internal motions of spiral nebulae also saw an abundance of care, if not a sufficiency. Measurements were precisely quantitative. Ranges of error were estimated and averages taken. Observational problems, such as that leading to a magnitude error, were known and the possibility that they had intervened was explored, albeit more by Hubble than by van Maanen. Instrumental difficulties were examined thoroughly and ruled out. Measuring instruments were kept free of large temperature fluctuations. A variety of plates taken with a variety of telescopes at a variety of exposure times by various observers were then placed in a variety of measuring instruments in a variety of positions and measured by various scientists. Furthermore, a null result was obtained in the measurement of an object other than a spiral nebula. Good scientists did all they humanly could to keep out personal bias, to keep science objective. It was not enough.

The one thing they did not do was the now traditional double-blind experiment common in testing medical drugs. Some patients receive the drug and others a placebo, with neither the patients nor the examining doctors knowing which is which. Bias remains, but without a purchase point. In many sciences, however, it is not always possible to set up a double-blind procedure. Even should it withstand future historical scrutiny, the double-blind procedure is less a general savior of all science than an admission of the potential failure of objectivity.

The history of modern science displays an increasingly critical examination of observational procedures by scientists and major advances toward the goal of objectivity. It also reveals that the goal has yet to be reached and suggests that the goal, while increasingly approachable, may be ultimately unreachable.

Observational bias is inevitable but not acceptable. Science and scientists are expected to winnow out errors at an early stage. To the extent that they do not, the objectivity of science is compromised. Historical cases reveal unacceptably long lifetimes for some erroneous observations and exceptional circumstances in the early resolution of others. In the instance of the claimed motions in spiral nebulae, it is not at all clear that the error would have been detected in such timely manner had not the significance of the measurements clashed with belief. Neither doubts about van Maanen's procedures nor an inability to reproduce his alleged measurements demanded refutation and rejection of his claim. Rather, it was the incompatibility of his claim with belief in the island universe hypothesis that led other scientists to question van Maanen's reported results. And it was only when evidence of an entirely different kind convincingly established the extragalactic nature of the spiral nebulae that van Maanen's alleged motions were quietly forgotten—never decisively explained as an error, the result of a personal bias.

As bedeviled by preconceptions as scientific observations are, their validity is not to be found completely in the process of discovery. Also useful is the attempt to validate scientific observation through a process of verification after the purported discovery. Such a process depends crucially upon the unrelenting rigor of peer review, however, and historical cases raise serious doubts about the effectiveness of peer review. Peer pressure seems more effective in silencing dissent, or at least in limiting controversy so that it does not reach public knowledge and bring science into public disrepute. In some cases the level of controversy has been kept low enough to protect the reputations of colleagues even within the scientific community.

At least one astronomer noticed the discrepancy between the Neptunian rings reported by Lassell and Challis, but he notified few others. A few more astronomers noticed the discrepancy between the directions of rotation of spiral nebulae reported by Slipher and by van Maanen, but they also severely limited the circle of scientists to whom they reported the discrepancy. Nor were van Maanen's improper claims of corroboration challenged in a public arena.

There is an obvious reluctance to throw the first stone, especially by residents of glass houses. If a young scientist does not realize this fact of life, a more senior member of the scientific community can instruct him or her in the culture of the group. In the case of van Maanen's alleged rotations, Lundmark was initially critical of van Maanen in print. Shapley was quick to inform Lundmark that while he (Shapley) was solely interested in getting at the truth and did not object to just criticism of his own work (somewhat linked to van Maanen's), there was little to gain if astronomers picked to pieces small and irrelevant points; Lundmark might well think about how many flaws or hasty conclusions someone might find in his big paper on the distances of globular clusters. Shapley had heard Russell's private criticism of a paper by Lundmark and was sure that Lundmark would not like to hear it, however much Russell generally admired Lundmark's work. That scientific theories are so sacrosanct that one can determine which anomalies are irrelevant is a disturbing implication of Shapley's letter. The letter silenced Lundmark.

More difficult to silence was Hubble, especially after a decade of effort had reinforced his disdain for van Maanen's work and that work remained the outstanding discrepancy in an otherwise consistent picture of the universe based largely on Hubble's own research. The Mount Wilson bureaucracy did not welcome a public dispute between two of its members, as letters from Seares to Hale show. Hubble was not to be completely silenced, but his eventual public statement was greatly muted relative to his many unpublished manuscripts.

There are also pressures on historians of recent science not to be too critical. When initial studies suggested unpleasant conclusions to come, there was talk by a few astronomers of closing archives to historians. And it may be easier to obtain research grants for more traditional investigations. When the results of studies have been critical of the scientific community or of individual scientists, some scientists have taken the studies almost as a personal attack and have responded with questions of the author's competence — typically in private, not in public.

Another obstacle to effective peer review occurs when observational data is regarded as the private property of the scientist. Such was the case with van Maanen's work on the rotation of spiral nebulae, and Hubble's investigation of the matter was slowed by the necessity to take new plates. Hubble allowed other astronomers free access to his plates.

Peer review could be more effective were data more freely available. It might also be more effective in a climate of opinion that encouraged vigilance by colleagues and competitors. A changing temper in the scientific community, with reputation coming to count for less, would be a promising sign for the progress of science.

Less than intense peer review devoted to winnowing out observational errors is mirrored by an excess of effort devoted to explaining away revealed errors as anything but the result of personal bias, as anything but a suspicion cast upon the supposed objectivity of science. The public explanation for Adams's error in the case of a gravitational redshift for Sirius B parallels that offered up in attempted exoneration of van Maanen and his internal motions in spiral nebulae. It parallels it in the scientists' unwillingness to admit the obvious effect of a personal bias, in the transparent inadequacy of the alternative explanation, and in the silent acquiescence of the scientific community to a patently fatuous alternative explanation.

Adams's alleged measurement of a gravitational redshift was contradicted by a later measurement at the Mount Wilson Observatory, and a passing attempt was made at explaining away both Adams's measurement and Moore's confirming measurement. The attempted explanation is, however, scarcely to be taken any more seriously than the measurements it attempts to explain away, except for what the error and the erroneous explanation reveal about the nature of scientists and science.

Scientists essentially have chosen to ignore earlier conflicting measurements. This strategy can be amply justified on practical grounds: it is unlikely that the earlier observations, however erroneous, can easily be ascribed convincingly to an instrumental error; and to the extent that a more powerful telescope promises better observations, the more modern measurement is to be preferred if given a choice between two otherwise equal but conflicting claims.

Nonetheless, as practical as the strategy at first appears, the casual dismissal of earlier observations for little reason other than their age and disagreement with more recent observations has revolutionary implications for science. Shakespeare said it well in *Richard II,* in the person of the Bishop of Carlisle protesting Bolingbroke's usurpation of Richard's crown (Act 4, scene 1):

> What subject can give sentence on his king? . . .
> And if you crown him, let me prophesy, —
> The blood of English shall manure the ground,
>
> And future ages groan for this foul act. . . .
> O, if you rear this house against this house,
> It will the woefullest division prove
> That ever fell upon this cursed earth:
> Prevent it, reject it, and let it not be so,
> Lest child, child's children, say against you —
> woe!

It was such a civil war within science that Shapley dissuaded Lundmark from waging and Mount Wilson officials dissuaded Hubble from waging. The larger issue, though, is legitimacy. Henry's future legal hold and that of his heirs on the throne could be no stronger than Richard's barring proof of the latter's illegitimate birth, not contemplated here. And even that sort of justification would have established a precedent always threatening future kings and the civil order. So it is, too, in science. Today's observations have the legitimacy and force of yesterday's but no more, at least until yesterday's are clearly shown to have had an illegitimate origin in instrumental error. The dismissal of observations on the charge of personal bias also has revolutionary implications for science. Usurpation of the place formerly held by what are now sentenced as errors due to some personal bias prophesies future tumultuous disorder. Observation against observation will the woefullest division prove.

However much one may yearn for stability in science, it is to be found no more in a blind allegiance to observation than political stability was found in the divine right of kings. Political progress has occurred, albeit haltingly, by expanding the decision-making and ratification base. In science, too, in practice if not yet in articulated rules and regulations, theory has joined observation in an uneasy coalition. It is time to recognize the existence of the coalition and to attempt to describe formally its reign.

Yet the refusal to admit that personal bias has penetrated

science lingers. No more plausible than Greenstein's purported explanation of the spurious gravitational redshift are the explanations put forth publicly attributing the alleged measurement of internal motions in spiral nebulae to anything but personal bias. Walter Baade, who had worked with Hubble at Mount Wilson in the 1930s to measure plates for possible motion and to show that there was no photographically detectable motion, later ascribed van Maanen's results to a magnitude error.[9] Baade could have been aware of Hubble's conclusion, never published, that there was evidence of a magnitude error. Baade might also be expected to have been familiar with Hubble's conclusion that any source of random error, anything other than a personal bias, was inadequate to account for the same direction of rotation for all seven spirals when their orientations on the photographic plates were random. And if Baade had not heard this argument from Hubble, he could have thought of it himself easily enough. Thus a question occurs about the extent to which Baade believed in the explanation which he offered. Would it have survived second thoughts? In fairness to Baade, it should be noted that his book was a posthumous publication, edited by another astronomer.

Another search for an explanation of van Maanen's spurious rotation of spiral nebulae was made in the 1970s by Richard Hart, then a graduate student. He objected to the statement that van Maanen had simply found what he wanted to find, and Hart devoted a good part of his doctoral dissertation to an examination of other possible sources of error. These he eliminated, and the last paragraph of the abstract of his dissertation began "Since only personal sources of error remain, a further computer simulation is performed in an attempt to explain a reasonable way in which such an error could occur." "Reasonable" apparently excludes reading one's expectations or wishes into the data. Hart continued, "It is found that a systematic displacement error in measuring plates with a stereocomparator can account for van Maanen's spurious results if the magnitude of the error is on the order of 0.002 mm (i.e., on the order of the accuracy attainable with the stereocomparator). . . ." Hart presented a plausible source of error imbedded in the stereocomparator and not dependent on a personal bias. The facts of the case, however, forced him to conclude the above sentence with a qualifying clause, "and if it is consistent with the orientation of the particular spiral features of the plate."[10] This, of course, is the catch that thwarted Hubble in his lengthy and repeated attempts to attribute van Maanen's spurious measurements to a magnitude error. Hart, in the body of his dissertation, noted:

> The conclusion, then, was that a simple error of measuring all points to one side cannot account for the rotational motions. To account for them a separate error displacement must be chosen for each point with the requirement that the net effect of such errors be systematic—that they contribute to a motion of the spiral as a whole.[11]

Though Hart ultimately could find no plausible source of error other than personal bias, summaries of his work have not always distinguished as sharply as they might between personal bias and instrumental error. One potentially ambiguous summary reads:

> When his [van Maanen's] photographic plates were re-checked, a computer analysis suggested that he had made perceptual errors of minute proportions (about 0.002 mm) which—because they were systematically in favor of his expectations—yielded an aggregate distortion large enough to retard the acceptance of a theory.[12]

This summary parallels Hart's dissertation abstract without emphasizing as much as is warranted that only a personal bias can explain van Maanen's results. Neither the use of a computer to analyze the error nor the minute size of the error invalidates the fact that the error was the product of expectation, the product of personal bias. Attempts to reject personal bias add something to an understanding of scientific measurement and much to an appreciation of just how precious is the idol of objective science.

Instances in which nonobjective influences have bent the course of scientific observation raise questions for scientists of how to conduct their business better in the future. Critiques of science can have constructive consequences.

All will welcome a more rigorous science. Some, if not all, will welcome a more realistic appraisal of the standing of science among human endeavors. And some, unfortunately, may delight in a downgrading of science beyond that suggested by reason, beyond that beneficial to the spiritual and material benefit of mankind.

Historical studies of science can pose a challenge to the high reputation which science currently commands. Science today is the queen of the intellectual disciplines, autonomous from other disciplines, with fundamental methods and data generally thought to be independent of other facets of our civilization. Indeed, the scientific method is emulated by other disciplines seeking an aura of respectability. There is no higher court of appeal than that of science; science is accepted by many as an ultimate value against which the worth of other intellectual disciplines is to be judged.

This situation has not always prevailed in the past, nor will it necessarily always prevail in the future. Science once was handmaiden to theology, the former but now deposed queen of the disciplines. In the medieval period in the Western world, theology was autonomous, with its own fundamental principles and methods. It provided people with knowledge considered to be of ultimate value.

The strength of science was an important determinant of whether science or theology would be victorious in the palace revolt, but the weakness of theology as revealed by historical studies also was an important factor. The divine sources of theology were found not to be absolute, but historically relative, impacted by cultural forces including the economic, the political, and the social.[13] The decline of the dominance of theology followed from historical studies that revealed the human nature and thus the human status of theology.

Historical studies which begin to investigate a possible human element of science similarly threaten to topple the current queen. It is in this context, as much or more so than for the improvement of the operations of science, that examination of the supposed objectivity of science may importantly influence future developments.

Notes

PREFACE

1. R. G. Collingwood, *The Idea of History* (Oxford: Oxford Univ. Press, 1946), 39; Plato, *Republic,* 361.
2. Fred Hoyle, "Steady State Cosmology Revisited," in *Cosmology and Astrophysics: Essays in Honor of Thomas Gold,* ed. Yervant Terzian and Elizabeth M. Bilson (Ithaca: Cornell Univ. Press, 1982), 17–57.
3. H. Bondi, "Fact and Inference in Theory and in Observation," *Vistas in Astronomy* 1(1955):155–62. Michael Crowe brought this article to my attention.
4. The article in question was Norriss S. Hetherington's "The Shapley-Curtis Debate," *Astronomical Society of the Pacific Leaflet* 490(1970):1–8.
5. Norriss S. Hetherington, "Lowell's Theory of Life on Mars," ibid. 501(1971):1–8.
6. J. Bronowski, *Science and Human Values,* rev. ed. (New York: Harper and Row, 1965), 29.

1. IMAGES OF SCIENCE

1. Henry James to Howard Sturgis, 5 August 1914, in *The Letters of Henry James,* vol. 2, ed. Percy Lubbock (New York: Scribner's, 1920), 382–85.
2. George Bernard Shaw, *Too True to be Good,* in *Complete Plays,* vol. 4 (New York: Dodd, Mead, 1962), 691–700.
3. Arnold Thackray and Robert K. Merton, "On Discipline Building: The Paradoxes of George Sarton," *Isis* 63(1972):473–95.
4. George Sarton, *Introduction to the History of Science,* vol. 1 (Baltimore: Williams and Wilkins, 1927), 3–4.
5. George Sarton, *The Study of the History of Science* (Cambridge: Harvard Univ. Press, 1936), 5.
6. Nicholas U. Mayall, "Edwin Powell Hubble, 1889–1953," *National Academy of Sciences Biographical Memoirs,* 41(1970):175–214.
7. Norman Campbell, *What is Science* (London: Methuen, 1921).

8. Norriss S. Hetherington, "Edwin Hubble: Legal Eagle," *Nature* 319(1986):189–90.

9. Edwin Hubble, manuscript on English Science in the Renaissance, 1937, Hubble Collection, box 2, Huntington Library, San Marino, California, hereafter cited as Hubble Collection, quoted in Norriss S. Hetherington, "Philosophical Values and Observation in Edwin Hubble's Choice of a Model of the Universe," *Historical Studies in the Physical Sciences* 13(1982):41–67.

10. Edwin Hubble, manuscript on Francis Bacon as a scientist, 1942, Hubble Collection.

11. Hetherington, "Philosophical Values," 41–67.

12. C. C. Gillispie, *The Edge of Objectivity: An Essay in the History of Scientific Ideas* (Princeton: Princeton Univ. Press, 1960).

13. George Magyar, "Pseudo-Effects in Experimental Physics: Some Notes for Case Studies," *Social Studies of Science* 7(1977):241–67; Mary Jo Nye, "N-rays: An Episode in the History and Psychology of Science," *Historical Studies in the Physical Sciences* 11(1980):125–56. See Pierre Quédec's "Weiss's Magneton: The Sin of Pride or a Venial Mistake?" submitted October 1986 to *Historical Studies in the Physical and Biological Sciences,* for a possibly similar instance, though Weiss may have reduced the dispersion in his results by careful omissions. Spencer Weart brought this paper to my attention.

14. Quoted in Horace F. Judson, *Search for Solutions* (New York: Holt, Rinehart and Winston, 1980), 22.

15. H. Bondi, "Fact and Inference in Theory and in Observation," *Vistas in Astronomy* 1(1955):155–62.

16. Norriss S. Hetherington, "Just How Objective Is Science," *Nature* 306(1983):727–30.

17. Derek J. de Solla Price, *Little Science, Big Science* (New York: Columbia Univ. Press, 1963), 11.

18. Stephen G. Brush, "Can Science Come Out of the Laboratory Now?" *Bulletin of the Atomic Scientists* 32(1976):40–43.

19. Keith Stewart Thomson, "The Sense of Discovery and Vice Versa," *American Scientist* 71(1983):522–24.

20. Henry Norris Russell, "Percival Lowell and His Work," *Outlook* 114(1916):781–83.

21. Steven Shapin, "History of Science and Its Sociological Reconstructions," *History of Science* 20(1982):157–211.

22. Here I am closely paraphrasing passages from Michael Mulkey, "Knowledge and Utility: Implications for the Sociology of Knowledge," *Social Studies of Science* 9(1979):63–80.

2. SEEING IS BELIEVING

1. John W. Shirley, ed., *Thomas Harriot, Renaissance Scientist* (Oxford: Clarendon Press, 1974); Shirley, *Thomas Harriot: A Biography* (Oxford: Clarendon Press, 1983).

2. Terrie F. Bloom, "Borrowed Perceptions: Harriot's Maps of the Moon," *Journal for the History of Astronomy* 9(1978):117–22. See also John Shirley, "Thomas Harriot's Lunar Observations," in *Science and History: Studies in Honor of Edward Rosen. Studia Copernicana XVI,* ed. Erna Hilfstein, Pawel Czartoryski, and Frank Grande (Wroclaw: Polish Academy of Sciences Press, 1978), 283–308.

3. Owen Gingerich, "Dissertatio cum Professore Righini et Sidereo Nuncio," in *Reason, Experiment, and Mysticism in the Scientific Revolution,* ed. M. L. Righini Bonelli and William R. Shea (New York: Science History Publications, 1975), 77–88.

4. Stillman Drake, *Galileo at Work: His Scientific Biography* (Chicago: Univ. of Chicago Press, 1978), 145, 154.

5. Bloom, "Borrowed Perceptions," 117–22.

6. J. D. North, "Thomas Harriot and the First Telescopic Observations of Sunspots," in *Thomas Harriot: Renaissance Scientist,* ed. Shirley, 129–65.

7. Ibid.

8. Stephen P. Rigaud, *Supplement to Dr. Bradley's Miscellaneous Works: With an Account of Harriot's Astronomical Papers* (Oxford: Oxford Univ. Press, 1833), 32. This may be found bound with Rigaud, *Miscellaneous Works and Correspondence of the Rev. James Bradley, D.D. F.R.S.* (Oxford: Oxford Univ. Press, 1832). See also Rigaud: "On Harriot's Papers," *Royal Institution of Great Britain Journal* 2(1831):267–71; and "On Harriot's Astronomical Observations contained in his unpublished Manuscripts belonging to the Earl of Egremont," *Proceedings of the Royal Society* 3(1832):125–26.

9. Nicholas Copernicus, *De Revolutionibus, Orbium Caelestium, Book 1, section 10;* translated in Thomas Kuhn, *The Copernican Revolution* (Cambridge: Harvard Univ. Press, 1957), 180; Norriss S. Hetherington, "The First Measurements of Stellar Parallax," *Annals of Science* 28(1972):19–25; Michael A. Hoskin, "Stellar Distances: Galileo's Method and Its Subsequent History," *Indian Journal of the History of Science* 1(1966):22–29; *Stellar Astronomy* (Chalfont St. Giles: Science History Publications, 1982), 5–11; J. D. Fernie, "The Historical Search for Stellar Parallax," *Journal of the Royal Astronomical Society of Canada* 69(1975):153–61 and 222–39.

10. Margaret 'Espinasse, *Robert Hooke* (Berkeley: Univ. of California Press, 1962), 3–5.

11. Robert Hooke, *An Attempt to Prove the Motion of the Earth by Observations* (London: 1674). This tract was reissued in Hooke, *Lectiones Cutterianae, or a Collection of Lectures: Physical, Mechanical,*

Geographical, and Astronomical. Made before the Royal Society on Several Occasions at Gresham College. To which are added divers Miscellaneous Discourses (London, 1679); facsimile reproduction in Robert W. T. Gunther, *Early Science in Oxford*, vol. 8 (Oxford, 1931). The tract is also available in the Landmarks of Science series. On Hooke's publications see Geoffrey L. Keynes, *A Bibliography of Dr. Robert Hooke* (Oxford: Clarendon Press, 1960).

12. References to Hooke in Thomas Birch's *The History of the Royal Society of London* (London, 1756 and New York: Johnson Reprint, 1968) and in William Derham's *Philosophical Experiments and Observations* (London, 1726) have been collected in Gunther, *Early Science in Oxford*. See Birch, *Royal Society* 2:96, 139, 313, 315; Gunther, *Early Science* 6:296, 289, 343.

13. Birch, *Royal Society* 2:394; Gunther, *Early Science* 6:354–55. Neither Birch nor Derham reports Hooke's observations; they are in Ms. Bradley 20, Bodleian Library, Oxford.

14. Birch, *Royal Society* 2:447; Gunther, *Early Science* 6:368.

15. Birch, *Royal Society* 2:448; Gunther, *Early Science* 6:369 and 279.

16. Gunther, *Early Science* 7:398, 399, 404, 406, 408–10, 416–17.

17. Hooke, *An Attempt to Prove the Motion of the Earth*, 24.

18. Ibid., 25.

19. Fernie, "The Historical Search for Stellar Parallax," 153–61 and 222–39.

20. Francis Baily, *An Account of the Revd. John Flamsteed, The First Astronomer Royal . . .* (London, 1835); and supplement (London, 1837). Flamsteed's observations were reported in John Wallis, *Opera Mathematica*, vol. 3 (Oxford, 1699), 701–8; reprinted in the Landmarks of Science series.

21. Jacques Ozanam, *Dictionaire Mathematique* (Amsterdam, 1691), 386.

22. Derham, *Philosophical Experiments and Observations*, 257–68.

23. James Bradley, "An Account of a new discovered Motion of the Fix'd Stars," *Philosophical Transactions of the Royal Society of London* 35(1729):637–61; Hoskin, *Stellar Astronomy*, 29–36.

24. Bradley, "An Account of a new discovered Motion," 637–61.

25. A. R. Hall and M. B. Hall, *The Correspondence of Henry Oldenburg*, vol. 5 (Madison: Univ. of Wisconsin Press, 1968), 112–17 and 241–46.

26. M. E. W. Williams, "Flamsteed's Alleged Measurement of Annual Parallax for the Pole Star," *Journal for the History of Astronomy* 10(1979):102–16. See also Williams, "James Bradley and the Eighteenth-Century 'Gap' in Attempts to Measure Annual Stellar Parallax," *Notes and Records of the Royal Society of London* 39(1982):83–100; K. M. Pedersen, "Roemer, Flamsteed and the Search for Stellar Parallax," *Vistas in Astronomy* 20(1976):105–9.

27. Williams, "Flamsteed's Alleged Measurement," 102–16.

3. PLANETARY FANTASIES: URANUS

1. William Herschel, *The Scientific Papers of Sir William Herschel* (London: Royal Society and Royal Astronomical Society, 1912), vol. 1, xvi. See also A. J. Turner, *Science and Music in 18th Century Bath* (Bath: Dawson and Goodall, 1977); Angus Armitage, *William Herschel* (London: Thomas Nelson, 1962); Roy Porter, "William Herschel, Bath, and the Philosophical Society," in *Uranus and the Outer Planets*, ed. Garry Hunt (Cambridge: Cambridge Univ. Press, 1982), 23–34.
2. William Herschel to Jacob Herschel, 12 April 1761; quoted in Turner, *Science and Music*, 24.
3. Herschel, *Scientific Papers*, xxix. See also Simon Schaffer, "Uranus and the Establishment of Herschel's Astronomy," *Journal for the History of Astronomy* 12(1981):11–26; J. A. Bennett, "Herschel's Scientific Apprenticeship and the Discovery of Uranus," in *Uranus and the Outer Planets*, ed. Hunt, 35–53; Robert W. Smith, "The Impact on Astronomy of the Discovery of Uranus," in *Uranus and the Outer Planets*, ed. Hunt, 81–89; A. F. O'D. Alexander, *The Planet Uranus* (New York: American Elsevier, 1965).
4. Herschel, *Scientific Papers*, xxx.
5. William Herschel, "Account of a Comet," *Philosophical Transactions of the Royal Society of London* 71(1781):492–501.
6. R. H. Austin, "Uranus Observed," *British Journal for the History of Science* 3(1967):275–84.
7. Thomas Hornsby to William Herschel, 26 February 1782, quoted in Schaffer, "Uranus and the Establishment of Herschel's Astronomy," 11–26.
8. Schaffer, "Uranus and the Establishment of Herschel's Astronomy," 11–26.
9. Ibid.
10. William Herschel, "On the Diameter and Magnitude of the Georgium Sidus; with a Description of the dark and lucid Disk and Periphery Micrometers," *Philosophical Transactions of the Royal Society of London* 73(1783):4–14.
11. Herschel, "Account of a Comet," 492–501.
12. Herschel, "On the Diameter and Magnitude," 4–14.
13. Schaffer, "Uranus and the Establishment of Herschel's Astronomy," 11–26.
14. Austin, "Uranus Observed," 275–84.
15. Ibid.

16. Ibid.
17. Herschel, "Account of a Comet," 492–501.
18. Austin, "Uranus Observed," 275–84.
19. William Herschel, "A Letter from William Herschel, Esq. F.R.S.," *Philosophical Transactions of the Royal Society of London* 73(1783):1–3.
20. Herschel, "On the Diameter and Magnitude," 4–14.
21. William Herschel, "On the Discovery of four additional Satellites of the Georgium Sidus. The retrograde Motion of its Old Satellites announced; and the Cause of their Disappearance at certain Distances from the Planet explained," *Philosophical Transactions of the Royal Society of London* 88(1798):47–79.
22. Austin, "Uranus Observed," 275–84.
23. Alexander, *The Planet Uranus.*
24. Herschel, "On the Discovery of four additional Satellites," 47–79.
25. Ibid.
26. William Herschel, "An Account of the Discovery of Two Satellites revolving round the Georgian Planet," *Philosophical Transactions of the Royal Society of London* 77(1787):125–29; "On the Georgian Planet and its Satellites," ibid. 78(1788):364–78.
27. Herschel, "On the Discovery of four additional Satellites," 47–79.
28. Ibid.
29. Stillman Drake, *Galileo at Work: His Scientific Biography* (Chicago: Univ. of Chicago Press, 1978), 166.
30. Herschel, "On the Discovery of four additional Satellites," 47–79.
31. Ibid.
32. Ibid.
33. Ibid.
34. Ibid.
35. Richard Baum, *The Planets: Some Myths and Realities* (Newton Abbot: David and Charles, 1973), Chap. 5.

4. PLANETARY FANTASIES: NEPTUNE

1. J. Challis to the editor of the *Cambridge Chronicle,* October 1846; the letter is filed as number 12 of some 47 letters written between 30 June 1846 and 10 March 1847 relating to the discovery of Neptune, gathered together and indexed by Dr. David W. Dewhirst at the Cambridge Observatories Archives. Hereafter cited as Neptune Papers. On the discovery of Neptune see also G. B. Airy, "Account of some Circumstances historically connected with the Discovery of the Planet exterior to Uranus," *Memoirs of the Royal Astronomical Society* 16(1847):385–414; James Challis, "An Account of Observations un-

dertaken in search of the Planet discovered at Berlin in Sept. 23, 1846," *Memoirs of the Royal Astronomical Society* 16(1847):415-26; H. H. Turner, *Astronomical Discovery* (London: Edward Arnold, 1904; Berkeley: Univ. of California Press, 1963), Chap. 2; Morton Grosser, *The Discovery of Neptune* (Cambridge: Harvard Univ. Press, 1962); Robert W. Smith, "William Lassell and the Discovery of Neptune," *Journal for the History of Astronomy* 14(1983):30-32. Michael Crowe brought the problem of Neptune's purported ring to my attention.

2. J. R. Hind, "Discovery of LeVerrier's Planet," *London Times*, 1 October 1846, 8.

3. Grosser, *The Discovery of Neptune*. For Airy's and Challis's reports see also Note 1.

4. Norriss S. Hetherington, "The Earl of Rosse's Experiments on Reflecting Telescopes," *Journal of the British Astronomical Association* 87(1977):472-77.

5. Smith, "William Lassell and the Discovery of Neptune," 30-32.

6. J. F. W. Herschel to William Lassell, 1 October 1846, quoted in Robert W. Smith and Richard Baum, "William Lassell and the Ring of Neptune: A Case Study in Instrumental Failure," *Journal for the History of Astronomy* 15(1984):1-17.

7. William Lassell to J. F. W. Herschel, 12 October 1846, quoted in Smith and Baum, "William Lassell and the Ring of Neptune," 1-17.

8. Smith and Baum, "William Lassell and the Ring of Neptune," 1-17. See also Norriss S. Hetherington, "Neptune's Supposed Ring," *Journal of the British Astronomical Association* 90(1979):20-29; William G. Hoyt, "Reflections Concerning Neptune's 'Ring,' " *Sky and Telescope* 55(1978):284-85.

9. William Lassell, *Diary,* 3 October 1846, quoted in Smith and Baum, "William Lassell and the Ring of Neptune," 1-17.

10. Ibid., 10 October 1846.

11. William Lassell to the editor of *London Times,* 14 October 1846, quoted in Richard Baum, *The Planets: Some Myths and Realities* (Newton Abbot: David and Charles, 1973), 128.

12. Ibid.

13. Smith and Baum, "William Lassell and the Ring of Neptune," 1-17.

14. William Lassell to the editor of the *Astronomische Nachrichten* 25(1847):197-200.

15. William Lassell to the editor of the *Monthly Notices of the Royal Astronomical Society* 7(1847):167-68.

16. Baum, *The Planets,* 133-34.

17. J. R. Hind to J. Challis, 30 September 1846; no. 12, Neptune Papers.

18. J. R. Hind to the editor of the *Astronomische Nachrichten,* 25(1847):205. See also the *Monthly Notices of the Royal Astronomical Society* 7(1847):168. Hind's written report in the former is more ten-

tative and does not give the inclination of the ring. The inclination of 30 degrees apparently was stated at the 11 December 1846 meeting of the Royal Astronomical Society and thence reported in print.

19. J. Challis to the editor of the *Cambridge Chronicle*, 16 October 1846; Neptune Papers, no. 16. See also J. Challis, *Special Report of Proceedings in the Observatory Relative to the New Planet* (12 December 1846); a copy of this pamphlet is in the Cambridge Observatories Library.

20. J. Challis, *Second Report of Proceedings in the Cambridge Observatory relating to the new Planet (Neptune)* (22 March 1847); reprinted in *Astronomische Nachrichten* 25(1847):309–14.

21. Ibid.

22. William Lassell to J. Challis, 19 January 1847; Neptune Papers, no. 43.

23. Ibid.

24. Letter from J. Challis to the editor of the *Astronomische Nachrichten*, 25(1847):229; Challis, *Second Report*, 309–14.

25. Lassell to *London Times*, 14 October 1846, quoted in Baum, *The Planets*, 128; Lassell to the *Astronomische Nachrichten* 25(1847):197–200.

26. Hind to the *Astronomische Nachrichten* 25(1847):205.

27. Challis, *Second Report*, 309–14.

28. W. R. Dawes to J. Challis, 7 April 1847; Neptune Papers, no. 44.

29. W. R. Dawes to J. F. W. Herschel, 21 April 1847, quoted in Smith and Baum, "William Lassell and the Ring of Neptune," 1–17.

30. Dawes to Challis, Neptune Papers, no. 44.

31. Ibid.

32. Ibid.

33. Robert Smith to the author, 11 May 1984.

34. Hetherington, "Neptune's Supposed Ring," 20–29.

35. William Lassell to the editor of the *Astronomische Nachricten* 26(1847):165.

36. Draft of a letter by J. Challis, 3 May 1847, Cambridge Observatories Archives, quoted in Hetherington, "Neptune's Supposed Ring," 20–29.

37. William Lassell, "Observations of Neptune and his Satellite," *Monthly Notices of the Royal Astronomical Society* 12(1851):155.

38. William Lassell to the editor of *Monthly Notices of the Royal Astronomical Society* 13(1853), 36.

39. William Lassell, entry for 11 November 1852 in Lassell Papers, quoted in Smith and Baum, "William Lassell and the Ring of Neptune," 1–17.

40. Ibid.

41. Ibid., 15 December 1852.

42. William Lassell, "Description of a Machine for Polishing Specula &c.,

with Directions for its Use, together with Remarks upon the Art of Casting and Grinding Specula, and a Description of a Twenty-foot Newtonian Telescope equatoreally mounted at Starfield, near Liverpool," *Memoirs of the Royal Astronomical Society* 18(1850);1-20.

43. Smith and Baum, "William Lassell and the Ring of Neptune," 1-17.
44. G. B. Airy, "Substance of the lecture delivered by the Astronomer Royal on the large reflecting Telescopes of the Earl of Rosse and Mr. Lassell, at the last November meeting," *Monthly Notices of the Royal Astronomical Society* 9(1848):110-21.
45. Smith and Baum, "William Lassell and the Ring of Neptune," 1-17.
46. G. B. Airy, *Account of the Northumberland equatoreal and dome, attached to the Cambridge Observatory* (Cambridge, 1844), 5-6. David Dewhirst brought this book to my attention.
47. F. J. Hargreaves, "The Northumberland Telescope at Cambridge Observatory," *The Observatory* 60(1937):322-25. David Dewhirst brought this article to my attention.
48. Robert Smith to the author, 11 May 1984.

5. PLANETARY FANTASIES: MARS

1. William G. Hoyt, *Lowell and Mars* (Tucson: Univ. of Arizona Press, 1976), 26; A. Lawrence Lowell, *Biography of Percival Lowell* (New York: Macmillan, 1935), 60; Ferris Greenslet, *The Lowells and their Seven Worlds* (Boston: Houghton Mifflin, 1946), 358-59. On Schiaparelli's eyesight, see Michael J. Crowe, *The Extraterrestrial Life Debate 1750-1900* (Cambridge: Cambridge Univ. Press, 1986), 496, 507.
2. Crowe, *Extraterrestrial Life*, 484-85.
3. George E. Webb, "The Planet Mars and Science in Victorian America," *Journal of American Culture* 3(1980):573-80; Hoyt, *Lowell and Mars*, Chap. 5.
4. W. W. Campbell, "*Mars*, by Percival Lowell," *Publications of the Astronomical Society of the Pacific* 8(1896):207-20.
5. Lowell's book *Mars* appeared in serial form in the *Atlantic Monthly* 75(1895):594-603 and 749-58; 76(1895):106-19 and 223-35; Percival Lowell, *Mars* (Boston: Houghton, Mifflin, 1895).
6. Norriss S. Hetherington, "Lowell's Theory of Life on Mars," *Astronomical Society of the Pacific Leaflet* 501(1971):1-8.
7. Mark Wicks, *To Mars Via the Moon* (Philadelphia: J. B. Lippincott, 1911).
8. Norriss S. Hetherington, "Amateur versus Professional: The British Astronomical Association and the Controversy over Canals on Mars," *Journal of the British Astronomical Association* 86(1976):303-8.

9. Norriss S. Hetherington, "Percival Lowell: Professional Scientist or Interloper?" *Journal of the History of Ideas* 42(1981):159–61.
10. Henry Norris Russell, "Percival Lowell and His Work," *Outlook* 114(1916):781–83.
11. Campbell, *"Mars, by Percival Lowell,"* 207–20.
12. Lowell, *Mars.*
13. Hoyt, *Lowell and Mars,* 325.
14. Ibid.
15. Ibid.
16. Percival Lowell, *Mars as the Abode of Life* (New York: Macmillan, 1908), 146–47.
17. Hoyt, *Lowell and Mars,* 65.
18. A. E. Douglass, "Double Canals and Canals in the Dark Regions," *Annals of the Lowell Observatory* 2, Chap. 5, Part 2 (1900):441–54.
19. Ibid.
20. Percival Lowell to W. D. McPherson, 12 June 1907, cited in Hoyt, *Lowell and Mars,* 123.
21. A. E. Douglass to W. L. Putnam, 12 March 1901, ibid., 124. See also G. E. Webb, "A. E. Douglass and the Canals of Mars," *Astronomy Quarterly* 3(1979):27–37; and *Tree Rings and Telescopes: The Scientific Career of A. E. Douglass* (Tucson: Univ. of Arizona Press, 1983).
22. Douglass to J. Jastrow, 9 January 1901, cited in Hoyt, *Lowell and Mars,* 124. See also A. E. Douglass, "Illusions of Vision and the Canals of Mars," *Popular Science Monthly* 70(1907):464–74.
23. Percival Lowell, "Width of the Double Canals of Mars with Different Apertures," *Lowell Observatory Bulletin* 5(1903):25–29.
24. Percival Lowell, "Experiment on the Visibility of fine lines in its bearings on the breadth of the 'canals' of Mars," ibid. 2(1903):1–4.
25. C. O. Lampland, "Notes on Visual Experiment," ibid. 10(1904):53–55.
26. J. E. Evans and E. Walter Maunder, "Experiments as to the Actuality of the 'Canals' observed on Mars," *Monthly Notices of the Royal Astronomical Society* 53(1903):488–99.
27. Percival Lowell, "Double Canals of Mars in 1903," *Lowell Observatory Bulletin* 15(1905):97–110.
28. Simon Newcomb, "The Optical and Psychological Principles Involved in the Interpretation of the So-Called Canals of *Mars,*" *Astrophysical Journal* 26(1907):1–17.
29. Ibid.
30. Percival Lowell, "The Canals of Mars, Optically and Psychologically Considered. A Reply to Professor Newcomb," ibid., 131–40.
31. Simon Newcomb, "Note on Preceding Paper," ibid., 141; Percival Lowell, "Reply to Professor Newcomb's Note," ibid., 142.
32. Hoyt, *Lowell and Mars,* 339.
33. Ibid.

34. Campbell, *"Mars, by Percival Lowell,"* 207–20.
35. Russell, "Percival Lowell and his Work," 781–83.
36. Eliot Blackwelder, "Mars as the Abode of Life," *Science* 29(1909):659–61.
37. F. R. Moulton, "A Reply to Dr. Percival Lowell," *Science* 30(1909):639–42.
38. Lowell, *Mars.*
39. Hetherington, "Lowell's Theory of Life on Mars," 1–8.
40. Ibid.
41. Ibid.

6. SIRIUS B AND THE GRAVITATION REDSHIFT

1. F. W. Bessel, "Extract from the translation of a Letter from Professor Bessell, On the Variations of the Proper Motions of *Procyon* and *Sirius.* Communicated by Sir. J. F. W. Herschel," *Monthly Notices of the Royal Astronomical Society* 6(1844):136–41; "Ueber Veränderlichkeit der eigener Bewegungen der Fixsterne," *Astronomische Nachrichten* 22(1844):145–90. See also Norriss S. Hetherington, "Sirius B and the Gravitational Redshift: An Historical Review," *Quarterly Journal of the Royal Astronomical Society* 21(1980):246–52.
2. C. A. F. Peters, "Ueber die eigene Bewegung des Sirius," *Astronomische Nachrichten* 32(1851):1–58; A. J. G. Auwers, "On the Irregularity of the Proper Motion of Sirius, and on a Missing Nebula," *Monthly Notices of the Royal Astronomical Society* 22(1862):148–50; T. H. Safford, "On the Proper Motion of Sirius in Declination," *Monthly Notices of the Royal Astronomical Society* 22(1862):145–48, and "On the Observed Motions of the Companion of Sirius," *Proceedings of the American Academy of Arts and Sciences* 6(1863):143–46.
3. D. J. Warner, *Alvan Clark and Sons, Artists in Optics* (Washington, D.C.: Smithsonian Institution Press, 1968).
4. Simon Newcomb, "New Refracting Telescope of the National Observatory, Washington, D.C.," in *The Science Record for 1874,* ed. Alfred E. Beach (New York: Munn, 1874), 324–32; G. P. Bond, "Discovery of a Companion of Sirius," *Monthly Notices of the Royal Astronomical Society* 22(1862):170; "On the Companion of Sirius," *Astronomische Nachrichten* 57(1862):131–34; "On the Companion of Sirius," *American Journal of Science and Arts* 33(1862):286–87. See also B. Z. Jones and L. G. Boyd, *The Harvard College Observatory: The First Four Directorships, 1839–1919* (Cambridge: Harvard Univ. Press, 1971), 123.
5. William G. Hoyt, *Lowell and Mars* (Tucson: Univ. of Arizona Press, 1976), 106.

6. W. S. Adams, "The Spectrum of the Companion of Sirius," *Publications of the Astronomical Society of the Pacific* 27(1915):236-37.

7. A. Einstein, "Über den Einflus der Schwerkraft auf die Ausbreitung des Lichtes," *Annalen der Physik* 35(1911):898-908.

8. L. E. Jewell, "The Coincidence of Solar and Metallic Lines," *Astrophysical Journal* 3(1896):89-113; H. Buisson and C. Fabry, "Measures de Pétites Variations de Longuers d'Onde Par la Méthode interférentielle. Application à Différents Problèmes de Spectroscopie Solaire," *Journal de Physique* 9(1910):298-316; J. Halm, "Über eine bisher unbekannte Verschiebung der Franuhoferschen Linien des Sonnenspektrums," *Astronomische Nachrichten* 173(1907):273-88; G. E. Hale and W. S. Adams, "A Photographic Comparison of the Spectra of the Limb and the Center of the Sun," *Astrophysical Journal* 25(1907):300-310. See also E. G. Forbes, "A History of the Solar Red Shift Problem," *Annals of Science* 17(1961):129-64.

9. E. Freundlich, "On the Empirical Foundation of the General Theory of Relativity," *Vistas in Astronomy* 1(1955):239-46.

10. Ronald W. Clark, *Einstein: The Life and Times* (New York: World, 1971), 174-76.

11. J. Evershed, "A New Interpretation of the General Displacement of the Lines of the Solar Spectrum Towards the Red," *Kodaikanal Observatory Bulletin* 3(1913):45-53; "Recollections of Seventy Years of Scientific Work," *Vistas in Astronomy* 1(1955):33-40; T. Royds, "A Preliminary Note on the Displacement to the Violet of Some Lines in the Solar Spectrum," *Kodaikanal Observatory Bulletin* 3(1914):59-69.

12. E. Freundlich, "Über die Verschiebung der Sonnenlinien nach dem roten Ende auf Grund der Hypothesen von Einstein und Nordström," *Physikalische Zeitschrift* 15(1914):369-71; "Über die Verschiebung der Sonnenlinien nach dem roten Ende des Spektrums auf Grund der Äquivalenzhypothese von Einstein," *Astronomische Nachrichten* 198(1914):265-70.

13. J. Evershed and T. Royds, "On the Displacements of the Spectrum Lines at the Sun's Limb," *Kodaikanal Observatory Bulletin* 3(1914):71-81.

14. J. Evershed, "The Displacement of the Cyanogen Bands in the Solar Spectrum," *The Observatory* 41(1918):371-75.

15. K. Schwarzchild, "Über die Verschiebungen der Bande bei 3883 $\overset{\circ}{A}$ im Sonnenspektrum," *Sitzungsberichte der Königlich Preussichen Akademie der Wissenschaften* 47(1914):1201-13; C. E. St. John, "The Principle of Generalized Relativity and the Displacement of Fraunhofer Lines Toward the Red," *Astrophysical Journal* 46(1917):249-65; L. Grebe and A. Bachem, "Über dem Einsteineffekt im Gravitationsfeld der Sonne," *Verhandlungen der Deutschen Physikalischen Gesellschaft* 21(1919):454-64, and "Über die Einstein-

verschiebung im Gravitationsfeld der Sonne," *Zeitschrift für Physik* 1(1920):51–54.

16. Arthur Eddington, *Space, Time, and Gravitation* (Cambridge: Cambridge Univ. Press, 1920), 130. See also F. W. Dyson, "On the Opportunity afforded by the Eclipse of 1919 May 29 of verifying Einstein's Theory of Gravitation," *Monthly Notices of the Royal Astronomical Society* 78(1917):445–47; A. S. Eddington, "The Total Eclipse of 1919 May 29 and the Influence of Gravitation on Light," *The Observatory* 42(1919):119–22, "Joint Eclipse Meeting of the Royal Astronomical Society," *The Observatory* 42(1919):389–98.

17. A. S. Eddington, "On the Relation between the Masses and Luminosities of the Stars," *Monthly Notices of the Royal Astronomical Society* 84(1924):308–32.

18. A. Vibert Douglas, *The Life of Arthur Stanley Eddington* (London: Thomas Nelson, 1956):73–74.

19. A. S. Eddington to W. S. Adams, early 1924, quoted in Douglas, *Life of Eddington,* 75.

20. Ibid.

21. W. S. Adams to A. S. Eddington, 12 February and 2 March 1924, ibid., 75–76.

22. A. S. Eddington to W. S. Adams, 22 March 1924, ibid., 76–77.

23. W. S. Adams, "The Relativity Displacement of the Spectral Lines in the Companion of Sirius," *Proceedings of the National Academy of Sciences* 11(1925):382–87, reprinted in *The Observatory* 48(1925):337–42; "A Study of the Gravitational Displacement of the Spectral Lines in the Companion of Sirius," *Publications of the Astronomical Society of the Pacific* 37(1925):158; "The Radial Velocity of the Companion of Sirius," *The Observatory* 49(1926):88; G. Strömberg, "Note Concerning the Radial Velocity of the Companion of Sirius," *Publications of the Astronomical Society of the Pacific* 38(1926):44.

24. Douglas, *Life of Eddington,* 77.

25. A. S. Eddington, *Stars and Atoms* (Oxford: Clarendon Press, 1927), 53. Ray Lyttleton brought this passage and the whole problem of Sirius B and the gravitational redshift to my attention.

26. Ibid., 51.

27. W. H. McCrea, "Einstein: Relations with the Royal Society," *Quarterly Journal of the Royal Astronomical Society* 20(1979):251–60.

28. Ibid.

29. Ibid. On Mercury see N. T. Roseveare, *Mercury's Perihelion from Le Verrier to Einstein* (Oxford: Clarendon Press, 1982).

30. J. H. Moore, "Recent Spectrographic Observations of the Companion of Sirius," *Publications of the Astronomical Society of the Pacific* 40(1928):229–33.

31. H. Bondi, "Fact and Inference in Theory and in Observation," *Vistas in Astronomy* 1(1955):155–62.

32. J. L. Greenstein, J. B. Oke, and H. L. Shipman, "Effective Temperature, Radius, and Gravitational Redshift of Sirius B," *Astrophysical Journal* 169(1971):563–66. See also J. L. Greenstein and V. L. Trimble, "The Einstein Redshift in White Dwarfs," *Astrophysical Journal* 149(1967):283–98; "The Gravitational Redshift of 40 Eridani B," *Astrophysical Journal* 175(1972):L1–L5; "The Einstein Redshift in White Dwarfs. III," *Astrophysical Journal* 177(1972):441–52; D. M. Popper, "Red Shift in the Spectrum of 40 Eridani B," *Astrophysical Journal* 120(1954):316–21.

33. Greenstein, Oke, and Shipman, "Effective Temperature," 316–21.

34. Ibid.

7. A GENERAL SOLAR MAGNETIC FIELD

1. George Ellery Hale, Biographical Notes, Hale Papers, California Institute of Technology, quoted in "George Ellery Hale 1898–1938," in *The Legacy of George Ellery Hale,* ed. Helen Wright, Joan N. Warnow, and Charles Weiner (Cambridge: MIT Press, 1972), i–iii. See also Helen Wright, *Explorer of the Universe. A Biography of George Ellery Hale* (New York: Sutton, 1966). For an introduction to the secondary literature concerning Hale and his milieu see Helen Wright, "Hale, George Ellery," *Dictionary of Scientific Biography* 6, 26–34.

2. George E. Hale, "A Study of the Conditions for Solar Research at Mount Wilson, California," *Astrophysical Journal* 21(1905):124–50; "The Solar Observatory of the Carnegie Institution of Washington," *Astrophysical Journal* 21(1905):151–72. See also the series of reports by Hale in the *Carnegie Institution of Washington Yearbook.*

3. J. C. Kapteyn, "Die mittlere Geschwindigkeit der Sterne, die Quantität der Sonnenbewegung und die mittlere Parallaxe der Sterne von verschiedener Grösse," *Astronomische Nachrichten* 146(1898):97–114. Subsequent papers appeared primarily in the *Publications of the Astronomical Laboratory at Groningen.* A major late paper is J. C. Kapteyn and P. J. Rhijn, "On the Distribution of the Stars in Space Especially in the High Galactic Latitudes," *Astrophysical Journal* 52(1920):23–38. On Kapteyn's work see W. de Sitter, *Kosmos* (Cambridge: Harvard Univ. Press, 1932), Chap. 4; E. Robert Paul, *Seeliger, Kapteyn and the Rise of Statistical Astronomy* (Ph.D. diss., Indiana University, 1976); "Kapteyn and the Dutch Astronomical Community," *Journal for the History of Astronomy* 12(1981):77–94.

4. Adriaan van Maanen, *The Proper Motions of 1418 Stars in and Near*

the Clusters h and x Persei (Utrecht: J. van Boekhoven, 1911).

5. Frederick H. Seares, "Adriaan van Maanen, 1884–1946," *Publications of the Astronomical Society of the Pacific* 58(1946):89–103.

6. Ibid.

7. Norriss S. Hetherington, "Adriaan van Maanen and Internal Motions in Spiral Nebulae: A Historical Review," *Quarterly Journal of the Royal Astronomical Society* 13(1972):25–39; "Adriaan van Maanen's Measurements of Solar Spectra for a General Magnetic Field," ibid. 16(1975):235–44.

8. George E. Hale, "The Heliomicrometer," *Astrophysical Journal* 25(1907):293–99; "Preliminary Note on the Rotation of the Sun as Determined from the Motions of the Hydrogen Flocculi," ibid. 27(1908):219–29; Walter S. Adams, "Spectrographic Observations of the Rotation of the Sun," *Astrophysical Journal* 26(1907):203–24.

9. On his early program of solar investigations see George E. Hale, "Solar Research at the Yerkes Observatory," *Astrophysical Journal* 16(1902):211–33; Hale and Walter S. Adams, "Photographic Observations of the Spectra of Sun-Spots," *Astrophysical Journal* 23(1906):11–44; Hale, Adams, and Henry G. Gale, "Preliminary Paper on the Cause of the Characteristic Phenomena of Sun-Spot Spectra," *Astrophysical Journal* 24(1906):185–13; Hale and Adams, "Second Paper on the Cause of the Characteristic Phenomena of Sun-Spot Spectra," *Astrophysical Journal* 25(1907):75–94; yearly reports by Hale in the *Carnegie Institution of Washington Yearbooks;* and Hale, "On the Probable Existence of a Magnetic Field in Sun-Spots," *Astrophysical Journal* 28(1908):315–43.

10. Henry W. Rowland, "On the Magnetic Effect of Electric Convection," *American Journal of Science and Arts,* ser. 3, 15(1878):30–38.

11. George E. Hale, "Solar Vortices," *Astrophysical Journal* 28(1908):100–116, reprinted in *Publications of the Astronomical Society of the Pacific* 20(1908):203–20; a summary of the paper appeared in *Nature* 78(1908):368–69; Hale, "Sunspots as Magnets and the Periodic Reversal of Their Polarity," *Nature* 113(1924):105–12.

12. P. Zeeman, "Doublets and Triplets in the Spectrum Produced by External Magnetic Forces," *Philosophical Magazine and Journal of Science,* 5th ser., 44(1897):55–60 and 255–59; "Lignes doubles et triples dans le spectre, produites sous l'influences d'un champ magnètique extérieur," *Comptes Rendu hebdomadaries des Séances de l'Académie des Sciences* 124(1897):1444–45.

13. J. Norman Lockyer, "Spectroscopic Observations of the Sun," *Proceedings of the Royal Society of London* 15(1867):256–58; C. A. Young, "Catalogue of Bright Lines in the Spectrum of the Solar Atmosphere," *Nature* 7(1872):17–20; A. Cortie, "Bands Observed in the Spectra of Sun-Spots at Stonyhurst Observatory," *Monthly Notices of the Royal Astronomical Society* 47(1886):19–22; Hale, "Solar Research at the

Yerkes Observatory," 211–33; Walter M. Mitchell, "Reversals in the Spectra of Sun-Spots," *Astrophysical Journal* 19(1904):357–59; C. A. Young, "Photography of Sun-Spot Spectra," *Astronomy and Astrophysics* 12(1893):649–50.

14. George E. Hale, "On the Probable Existence of a Magnetic Field in Sun-Spots," *Astrophysical Journal* 28(1908):315–43; "Solar Vortices and the Zeeman Effect," *Publications of the Astronomical Society of the Pacific* 20(1908):220–24; "The Zeeman Effect in the Sun," *Publications of the Astronomical Society of the Pacific* 20(1908):287–88; "Solar Vortices," *Nature* 77(1908):368–69.

15. P. Zeeman, "Solar Magnetic Fields and Spectrum Analysis," *Nature* 78(1908):369–70.

16. T. G. Cowling, "Magnetic Fields in the Sun," *Quarterly Journal of the Royal Astronomical Society* 12(1971):348–51.

17. Hale, "Solar Vortices and the Zeeman Effect," 220–24, and "The Zeeman Effect in the Sun," 287–88; "Preliminary Results of an Attempt to Detect the General Magnetic Field of the Sun," *Astrophysical Journal* 38(1913):27–98; "Mount Wilson Solar Observatory," *Carnegie Institution of Washington Yearbook* 11(1912):172–213; 12(1913):195–240; 13(1914):241–48; 14(1915):251–93.

18. Hale, "Preliminary Results of an Attempt to Detect the General Magnetic Field of the Sun," 27–98.

19. Ibid.

20. Ibid.

21. F. H. Seares, A. van Maanen, and F. Ellerman, "The Location of the Sun's Magnetic Axis," *Proceedings of the National Academy of Sciences* 4(1918):4–9.

22. Seares, van Maanen, and Ellerman, "Location of Sun's Magnetic Axis," 4–9; "Deviations of the Sun's General Magnetic Field from that of a Uniformly Magnetized Sphere," *Proceedings of the National Academy of Sciences* 5(1919):242–46.

23. T. G. Cowling to author, 5 November 1975.

24. Ibid.

25. F. J. M. Stratton, "John Evershed 1864–1956," *Biographical Memoirs of Fellows of the Royal Society* 3(1957):41–51; J. Evershed, "Recollections of Seventy Years of Scientific Work," *Vistas in Astronomy* 1(1955):33–40.

26. J. Evershed, "A New Method of Using a Spectrograph for Solar Rotation Work," *Monthly Notices of the Royal Astronomical Society* 93(1933):165–68.

27. J. Evershed, "On the Detection of Small Doppler Shifts in the Spectrum of the Reversing Layer," *Monthly Notices of the Royal Astronomical Society* 94(1933):96–98.

28. J. Evershed, "Note on the Zeeman Effect in Sunspot Spectra," *Monthly Notices of the Royal Astronomical Society* 99(1939):217–18;

"Measures of the Relative Shifts in the Line 5250.218 and Neighboring Lines in Mt. Wilson Solar Magnetic Field Spectra," *Monthly Notices of the Royal Astronomical Society* 99(1939):438–40.

29. George E. Hale, Walter S. Adams, and Frederick H. Seares, "Mount Wilson Observatory: Survey of the Year's Work," *Carnegie Institution of Washington Yearbook* 33(1933–1934):125–57.

30. T. G. Cowling to author, 5 November 1975. See also "Meeting of the Royal Astronomical Society," *The Observatory* 69(1939):85–100.

31. Ibid.

32. George E. Hale, Walter S. Adams, and Frederick H. Seares, "Mount Wilson Observatory: Survey of the Year's Work," *Carnegie Institution of Washington Yearbook* 33(1933–1934):125–57.

33. R. M. Langer, "Current Investigations of the General Magnetic Field of the Sun," *Publications of the Astronomical Society of the Pacific* 48(1936):208–9; George E. Hale, "Solar Magnetism," *Nature* 136(1935):703–5.

34. H. W. Babcock and T. G. Cowling, "General Magnetic Fields in the Sun and Stars," *Monthly Notices of the Royal Astronomical Society* 113(1953):357–81.

35. Frederick H. Seares, "Adriaan van Maanen, 1884–1946," 89–103; Gerald P. Kuiper, "German Astronomy during the War," *Popular Astronomy* 54(1946):263–87. Kuiper's public assessment of German astronomers' cooperation with Nazi policies is somewhat at variance with what he wrote in private letters.

36. Georg Thiessen, "La mesure du champ magnétique général du Soleil," *Annales d'Astrophysique* 9(1946):101–19; "Über die Messung des allgemeinen Magnetfeldes der Sonne," *Forschungsbericht des Fraunhofer-Instituts* 8(1945); "Über eine empfindliche Anordung zur Messung Kleinster Zeeman-Verschiebungen im Sonnespektrum," *Die Himmelswelt* 55(1947):22–26.

37. T. G. Cowling to author, 5 November 1975. Thiessen had tried to compensate for Zeeman displacements by a superadded pressure pulsation, and as he increased his accuracy his pulsations became less and less; the first measures, with the interference rings pulsating with the reversal of polarity, were most impressive but turned out to be in error; G. Thiessen, "The Sun's Magnetic Field," *The Observatory* 69(1949):228; "Measurements of the Sun's General Magnetic Field," *Nature* 169(1952):147; "Lichtelektrische Messungen des solaren Magnetfeldes," *Zeitschrift für Astrophysik* 30(1952):185–99.

38. P. tenBruggencate and H. von Klüber, "Physik der Sonne," *FIAT Review of German Science, 1939–1946: Astronomy, Astrophysics and Cosmology* (Wiesbaden: Office of Military Government for Germany Field Information Agencies, Technical, 1948), 181–228; H. van Klüber, "An Attempt to Detect a General Magnetic Field of the Sun by a Spectrographic Method, Using a Lummer Plate," *Monthly Notices of*

the Royal Astronomical Society 111(1951):2-17; "Remarks on Hale's Determination of the General Magnetic Field of the Sun," *Monthly Notices of the Royal Astronomical Society* 112(1952):540-45; "Further Measurements to Detect a General Magnetic Field of the Sun," *Monthly Notices of the Royal Astronomical Society* 114(1954):242; "Spectroscopic Measurements of Magnetic Fields on the Sun," *Vistas in Astronomy* 1(1955):751-76.

39. Harold D. Babcock, "Recent Progress in the Study of the General Magnetic Field of the Sun," *Publications of the Astronomical Society of the Pacific* 53(1941):237-38; Ira S. Bowen, "Survey of the Year's Work at the Mount Wilson and Palomar Observatories," *Publications of the Astronomical Society of the Pacific* 61(1949):243-53; K. O. Kiepenheuer, "Photoelectric Measurements of Solar Magnetic Fields," *Astrophysical Journal* 117(1953):447-53; "Ist das allgemeine Magnetfeld der Sonne meßbar?" *Zeitschrift für Naturforschung* 8a(1953):225-27; Horace W. Babcock and Harold D. Babcock, "Mapping the Magnetic Fields of the Sun," *Publications of the Astronomical Society of the Pacific* 64(1952):282-87; Horace W. Babcock, "The Solar Magnetograph," *Astrophysical Journal* 118(1953):387-96; H. W. Babcock and T. G. Cowling, "General Magnetic Fields in the Sun and Stars," 357-81; Horace W. Babcock and Harold D. Babcock, "The Sun's Magnetic Field and Corpuscular Emission," *Nature* 175(1955):349-66; Harold D. Babcock, "The Sun's Polar Magnetic Field," *Astrophysical Journal* 130(1959):364-65; H. W. Babcock, "The Topology of the Sun's Magnetic Field," *Annual Review of Astronomy and Astrophysics* 1(1963):41-58.

40. H. W. Babcock and T. G. Cowling, "General Magnetic Fields in the Sun and Stars," 357-81; H. W. Babcock, "The Topology of the Sun's Magnetic Field," 41-58; A. Severny, "Solar Magnetic Fields," *Space Science Reviews* 3(1964):451-86.

41. Jan Olof Stenflo, "Hale's Attempt to Determine the Sun's General Magnetic Field," *Solar Physics* 14(1970):263-73. Sydney van den Berg brought this paper and the problem of general solar magnetic field to my attention.

42. Ibid.

43. T. G. Cowling to author, 5 November 1975.

44. Cowling to author, 5 November 1975; Harlow Shapley, "Henry Norris Russell Oct. 25, 1877-Feb. 18, 1957," *National Academy of Sciences Biographical Memoirs* 32(1958):354-78.

8. THE PURPORTED ROTATION OF SPIRAL NEBULAE

1. Norriss S. Hetherington, "Cosmic Perspective: Man, Society, and the Universe," *Mercury* 4(1975):9–13; Richard Berendzen, "Geocentric to Heliocentric to Galactocentric to Accentric: The Continuing Assault to the Egocentric," in *Copernicus, Yesterday and Today,* ed. A. Beer and K. Strand, *Vistas in Astronomy,* 17(1975): 65–83; F. R. Johnson and S. V. Larkey, "Thomas Digges, the Copernican System, and the Idea of the Infinity of the Universe in 1576," *Huntington Library Bulletin* 5(1934):69–117; F. R. Johnson, "The Influence of Thomas Digges on the Progress of Modern Astronomy in Sixteenth-Century England," *Osiris* 1(1936):390–410; *Astronomical Thought in Renaissance England* (Baltimore: Johns Hopkins Univ. Press, 1937); Arthur Lovejoy, *The Great Chain of Being* (Cambridge: Harvard Univ. Press, 1936); Grant McColley, "The Ross-Wilkins Controversy," *Annals of Science* 3(1938):153–89; Steven J. Dick, *Plurality of Worlds: The Origins of the Extra Terrestrial Life Debate from Democritus to Kant* (Cambridge: Cambridge Univ. Press, 1982); Marjorie Nicholson, "A World in the Moon," *Smith College Studies in Modern Languages* 172(1936):1–72.
2. William Whiston, *Astronomical lectures* (London: R. Senex and W. Taylor, 1715); see also Whiston's *Astronomical Principles of Religion, Natural, and Reveal'd* (London: J. Senex and W. Taylor, 1717), his autobiographical *Memoirs of the Life and Writings of Mr. William Whiston . . .* (London: Whiston, 1749), and Michael Hoskin, *Stellar Astronomy* (Chalfont St Giles: Science History Publications, 1982), 67–100. Thomas Wright, *An Original Theory or New Hypothesis of the Universe . . .* (London, 1750); on Wright's cosmology, see the introductions by Michael Hoskin to facsimile reprints of Wright's works and Hoskin's "The Cosmology of Thomas Wright of Durham," *Journal for the History of Astronomy,* 1(1970):44–52, and "The English Background to the Cosmology of Wright and Herschel," in *Cosmology, History, and Technology,* ed. Wolfgang Yourgrau and Allen D. Breck (New York: Plenum Press, 1977), 219–31; on Wright's life, see Edward Hughes, "The Early Journal of Thomas Wright of Durham," *Annals of Science,* 7(1951):1–24; and on the problem of the Milky Way, see Stanley L. Jaki, *The Milky Way: An Elusive Road for Science* (New York: Science History Publications, 1972). Immanuel Kant, *Allgemeine Naturgeschichte und Theorie des Himmels* (Konigsberg: J. F. Petersen, 1755), English translation by W. Hastie, *Kant's Cosmology* (Glasgow: J. Maclehose, 1900); Norriss S. Hetherington, "Sources of Kant's Model of the Stellar System," *Journal of the History of Ideas* 34(1973):461–62; Kenneth Glyn Jones, "The Observational Basis for Kant's *Cosmogony:* A critical analysis," *Journal for the History of Astronomy* 2(1971):23–34; and G. J. Whitrow, "Kant and the Extra Galactic Nebulae," *Quar-*

terly Journal of the Royal Astronomical Society 8(1967):48–56. P. L. M. de Maupertuis, "Discours sur les differentes figures des astres . . ." *Memoires de l'Académie royale des Sciences pour l'Annee MDCCXXXIV* (Paris, 1736), 78–83, translated in Kenneth Glyn Jones, "The Search for the Nebulae, Part 4," *Journal of the British Astronomical Association* 79(1968–69):19–25, and Jones, *The Search for the Nebulae* (Chalfont St. Giles: Alpha Academic/New York: Neale Watson Academic Publications, 1975). J. H. Lambert, *Cosmologischen Briefe Über die Einrichtung des Weltbanes* (Augsburg: Eberhard Kletts Wittib., 1761), translated in Stanley L. Jaki, *Cosmological Letters on the Arrangement of the World Edifice* (Edinburgh: Scottish Academic Press/New York: Science History Publications, 1976); J. H. Lambert to Immanuel Kant, 13 November 1765, in *Kant Philosophical Correspondence, 1759–99,* Arnulf Zweig, editor and translator (Chicago: Univ. of Chicago Press, 1967), 43–47.

3. J. L. E. Dreyer, *The Scientific Papers of Sir William Herschel* (London: The Royal Society and the Royal Astronomical Society, 1912); Michael A. Hoskin, *William Herschel and the Construction of the Heavens* (London: Oldbourne, 1963); William Herschel, "Account of Some Observations Tending to Investigate the Construction of the Heavens," *Philosophical Transactions of the Royal Society of London* 74(1784):437–51; "On the Construction of the Heavens," *Philosophical Transaction of the Royal Society of London* 75(1785):213–66; "Catalogue of a Second Thousand of New Nebulae and Clusters of Stars, with a few introductory Remarks on the Construction of the Heavens," *Philosophical Transactions of the Royal Society of London* 79(1789):212–26. See also Michael Hoskin, "Apparatus and Ideas in Mid-nineteenth-century Cosmology," *Vistas in Astronomy* 9(1967):79–85; this is a particularly important paper discussing not only the observations of Herschel, but also observations of Airy, Rosse, Robinson, Nichol, John Herschel, and de la Rue.

4. William Herschel, "On Nebulous Stars, properly so called," *Philosophical Transactions of the Royal Society of London* 81(1791):71–88.

5. Papers by the Earl of Rosse on the construction of reflecting telescopes are reprinted in Sir Charles Parsons, ed., *The Scientific Papers of William Parsons Third Earl of Rosse 1800–1867* (London: Percy Lund, Humphries, 1926); see also Norriss S. Hetherington, "The Earl of Rosse's Experiments on Reflecting Telescopes," *Journal of the British Astronomical Association* 87(1977):472–77. On resolving nebulae, see the Earl of Rosse, "Observations on some of the Nebulae," *Philosophical Transactions of the Royal Society of London* 134(1844):321–23; T. R. Robinson, "On Lord Rosse's Telescope," *Proceedings of the Royal Irish Academy* 3(1845):114–33; "On Lord

Rosse's Telescope," *Proceedings of the Royal Irish Academy* 4(1848):119. On the interpretation of Huggins's spectra, see Warren de la Rue, "The President's Address," *Monthly Notices of the Royal Astronomical Society* 25(1864):1–17; Daniel Kirkwood, "On the Testimony of the Spectroscope to the truth of the Nebular Hypothesis," *American Journal of Science*, ser. 2, 2(1871):155–56.

6. Agnes M. Clerke, *A Popular History of Astronomy*, 2d ed. (Edinburgh: Adam and Charles Black, 1887), 437–38.

7. V. M. Slipher, "The Radial Velocity of the Andromeda Nebula," *Lowell Observatory Bulletin* 2(1913):56–57; "The Detection of Nebular Rotation," *Lowell Observatory Bulletin* 2(1913):66; "Spectrographic Observations of Nebulae," *Popular Astronomy* 33(1915):21–24. See also R. W. Smith, "The Origins of the Velocity-Distance Relation," *Journal for the History of Astronomy* 10(1979):133–65; and Norriss S. Hetherington, "The Development and Early Application of the Velocity-Distance Relation" (Indiana Univ.: Ph.D. diss., 1970/Ann Arbor: University Microfilms International, 71-06860); "The Measurement of Radial Velocities of Spiral Nebulae," *Isis* 62(1971):309–13. On the debate over the nature of the nebulae, see Harlow Shapley, "On the Existence of External Galaxies," *Publications of the Astronomical Society of the Pacific* 31(1919):261–68. Three previous review articles are P. Puiseux, "Les nébuleuses spirales," *Revue Scientifique* 1(1912):417–22; A. C. D. Crommelin, "Are the Spiral Nebulae external Galaxies?" *Scientia* 21(1917):365–76 [reprinted in *Journal of the Royal Astronomical Society of Canada* 12(1917):33–46]; and Heber D. Curtis, "Modern theories of the Spiral nebulae," *Journal of the Washington Academy of Sciences* 9(1919):217–27. Edwin Hubble's early opinion is found in his "Photographic Investigations of Faint Nebulae," *Publications of the Yerkes Observatory* 4(1920):69–85; see also Norriss S. Hetherington, "Edwin Hubble and the Development of Modern Cosmology," *Proceedings No. 2, XIVth International Congress for the History of Science* (Tokyo: Science Council of Japan, 1975), 261–64.

8. Adriaan van Maanen, "Preliminary Evidence of Internal Motion in the Spiral Nebula Messier 101," *Astrophysical Journal* 44(1916):210–28.

9. Shapley, "On the Existence of External Galaxies," 261–68, and Shapley and H. D. Curtis, "The Scale of the Universe," *Bulletin of the National Research Council* 2(1921):171–217. On the Shapley-Curtis presentation, see Otto Struve, "A Historic Debate about the Universe," *Sky and Telescope* 19(1960):398–401; Norriss S. Hetherington, "The Shapley-Curtis Debate," *Astronomical Society of the Pacific Leaflet* 490(1970):1–8; Richard Berendzen, Richard Hart, and Daniel Seeley, *Man Discovers the Galaxies* (New York: Science History Publications, 1976); M. A. Hoskin, "The 'Great De-

bate': What Really Happened," *Journal for the History of Astronomy* 7(1976):169–82, and *Stellar Astronomy,* 175–88; and Robert W. Smith, "The Great Debate Revisited," *Sky and Telescope* 65(1983):28–29.

10. V. M. Slipher, "The Detection of Nebular Rotation," *Lowell Observatory Bulletin* 2(1914):66. See also Charles Fabry and H. Buisson, "Application of the Interface Method to the Study of Nebulae," *Astrophysical Journal* 33(1911):406–9; Buisson, Fabry, and H. Bourget, "An Application of Interference to the Study of the Orion Nebula," *Astrophysical Journal* 40(1914):241–58; Edwin B. Frost and Charles H. Maney, "New Conceptions of the Nebula of Orion," *Popular Astronomy* 23(1915):485–87; W. W. Campbell and J. H. Moore, "Observed Rotations of Planetary Nebulae," *Publications of the Astronomical Society of the Pacific* 27(1915):245–47; Edward C. Pickering, *Harvard College Observatory Bulletin* 591(1915); Francis G. Pease, "Radial Velocity of the Andromeda Nebula," *Publications of the Astronomical Society of the Pacific* 27(1915):134; Pease, "The Spiral Nebula Messier 33," *Publications of the Astronomical Society of the Pacific* 26(1916):33–34; M. Wolf, "Heidelberg," *Astronomische Gesellschaft, Viertel Jahreschrift* 49(1914):151–63.

11. Adriaan van Maanen, "Investigations on Proper Motion, Eleventh Paper: The Proper Motion of Messier 13 and Its Internal Motion," *Astrophysical Journal* 61(1925):130–36.

12. C. O. Lampland to Percival Lowell, 12 July 1916, *Early Correspondence of the Lowell Observatory 1894–1916* (Flagstaff, Ariz.: Lowell Observatory, 1973), roll 2, frame 655, microfilm edition; C. O. Lampland, "Preliminary Measures of the Spiral Nebulae N.G.C. 5194 (M91) and N.G.C. 4254 (M99) for Proper Motion and Rotation," *Popular Astronomy* 24(1916):667–68; "Nineteenth Meeting of the American Astronomical Society," *Popular Astronomy* 24(1916):578–85; Heber D. Curtis, "Descriptions of 132 Nebulae and Clusters Photographed with the Crossley Reflector," *Lick Observatory Bulletin* 7, No. 219(1912):81–84; "Descriptions of 109 Nebulae and Clusters Photographed with the Crossley Reflector: Second List," *Lick Observatory Bulletin* 8, No. 248(1913):43–46; "Preliminary Note on Nebular Proper Motions," *Proceedings of the National Academy of Sciences* 1(1915):10–12.

13. Norriss S. Hetherington, "The Simultaneous 'Discovery' of Internal Motions in Spiral Nebulae," *Journal for the History of Astronomy* 6(1975):115–25.

14. T. C. Chamberlin, "A Group of Hypotheses Bearing on Climatic Changes," *Journal of Geology* 5(1897):653–83; "An Attempt to Test the Nebular Hypothesis by the Relations of Masses and Momenta," *Journal of Geology* 8(1900):58–73; F. R. Moulton, "An Attempt to Test the Nebular Hypothesis by an Appeal to the Laws of Motion,"

Astrophysical Journal 11(1900):103-30; T. C. Chamberlin and F. R. Moulton, "Certain Recent Attempts to Test the Nebular Hypothesis," *Science* 12(1900):201-208. See also Stephen G. Brush, "A Geologist among Astronomers: The Rise and Fall of the Chamberlin-Moulton Cosmogony," *Journal for the History of Astronomy* 9(1978):1-41 and 77-104. For the pictures of spiral nebulae, see James E. Keeler, "The Crossley Reflector of the Lick Observatory," *Astrophysical Journal* 11(1900):325-49, and "Photographs of Nebulae and Clusters, Made with the Crossley Reflector, by James Edward Keeler, Director of the Lick Observatory, 1898-1900," *Publications of the Lick Observatory* 8(1908):1-46, 71. T. C. Chamberlin, "Fundamental Problems of Geology," *Carnegie Institution of Washington Yearbook* 3(1904):195-258; 14(1915):368; 15(1916):358-59; 16(1917):307-19.

15. Adriaan van Maanen, "Preliminary Evidence of Internal Motion in the Spiral Nebula Messier 101 from Plates taken by G. W. Ritchey, J. E. Keeler, C. D. Perrine and H. D. Curtis, measured and discussed by A. van Maanen," 13-page handwritten draft, Hale Observatories, Pasadena, California. See Norriss S. Hetherington, "Adriaan van Maanen on the Significance of Internal Motions in Spiral Nebulae," *Journal for the History of Astronomy* 5(1974):52-53.

16. Frederick H. Seares, "Adriaan van Maanen, 1884-1946," *Publications of the Astronomical Society of the Pacific* 58(1946):89-103; A. van Maanen, "The Relative Proper-Motions of 162 Stars in the Neighborhood of the Great Nebula in Orion," *Astronomical Journal* 27(1912):139-46; W. S. Adams and A. van Maanen, "A Group of Stars of Common Motion in the h and x Persei Clusters," *Astronomical Journal* 27(1912):187-88.

17. van Maanen, "Preliminary Evidence," 210-28.

18. Deborah J. Mills (Warner), "George Willis Ritchey and the Development of Celestial Photography," *American Scientist* 54(1966):64-93. See also G. W. Ritchey, "The Modern Photographic Telescope and the New Astronomical Photography," *Journal of the Royal Astronomical Society of Canada* 22(1928):159-77, 207-30, 303-24, 359-82; 23(1929):15-36, 167-90; "On the Modern Reflecting Telescope and the Making and Testing of Optical Mirrors," *Smithsonian Contributions to Knowledge* 34(1904):1-51; "Notes on Photographs of Nebulae made with the 60-Inch Reflector of the Mount Wilson Observatory, by Professor G. W. Ritchey, Assoc. R.A.S.," *Monthly Notices of the Royal Astronomical Society* 70(1910):623-25, 647-49.

19. Walter Baade, *Evolution of Stars and Galaxies* (Cambridge: Harvard Univ. Press, 1963), 28-29.

20. van Maanen, "Preliminary Evidence," 210-28.

21. van Maanen, "The Relative Proper-Motions of 162 Stars," 139-46.

22. Edwin Hubble, untitled, undated, handwritten note on the nature of stellar and nebular images on photographic plates, box 2, Hubble Collection, Huntington Library, San Marino, California; untitled; undated, handwritten note on the possibility of a magnitude error in van Maanen's work, box 4, Hubble Collection; "Angular Rotations of Spiral Nebulae," unpublished, undated manuscript, box 4, Hubble Collection; Norriss S. Hetherington, "Edwin Hubble on Adriaan van Maanen's Internal Motions in Spiral Nebulae," *Isis* 65(1974):390-93; "Edwin Hubble's Examination of Internal Motions in Spiral Nebulae," *Quarterly Journal of the Royal Astronomical Society* 15(1974):392-418.

23. Hetherington, "Hubble's Examination of Internal Motions," 392-418; J. D. Fernie, "The Historical Quest for the Nature of the Spiral Nebulae," *Publications of the Astronomical Society of the Pacific* 82(1970):1189-1230; Norriss S. Hetherington, "Adriaan van Maanen and Internal Motions in Spiral Nebulae: A Historical Review," *Quarterly Journal of the Royal Astronomical Society* 13(1972):25-39.

24. van Maanen, "Preliminary Evidence," 210-28.

25. George E. Hale, "Mount Wilson Solar Observatory," *Carnegie Institution of Washington Yearbook* 15(1916):227-72.

26. *Carnegie Institution of Washington Yearbook* 15(1916):227-72; 16(1917):199-235; 17(1918):181-218; 18(1919):217-64.

27. *Carnegie Institution of Washington Yearbook* 19(1920):209-65; Adriaan van Maanen, "Internal Motion in the Spiral Nebula Messier 33," *Proceedings of the National Academy of Sciences* 7(1921):1-5.

28. F. G. Pease, "Radial Velocity of the Andromeda Nebula," *Publications of the Astronomical Society of the Pacific* 27(1915):134; "The Rotation and Radial Velocity of the Central Part of the Andromeda Nebula," *Proceedings of the National Academy of Sciences* 4(1918):21-24.

29. C. O. Lampland, "Preliminary Measures of the Spiral Nebulae N.G.C. 5194 (M51) and N.G.C. 4254 (M99) for Proper Motion and Rotation," *Popular Astronomy* 24(1916):667-68, reprinted in *Publications of the American Astronomical Society* 3(1918):206-7.

30. W. J. A. Schouten, "Probable Motions in the Spiral Nebula Messier 51 (Canes Venatici)," *The Observatory* 42(1919):441-44.

31. S. Kostinsky, "Probable Motions in the Spiral Nebula Messier 51 (Canes Venatici) found with the Stereo-comparator. Preliminary Communication," *Monthly Notices of the Royal Astronomical Society* 77(1917):233-34.

32. E. A. Kreiken, "On the Differential Measurement of Proper Motion," *The Observatory* 43(1920):255-60.

33. Adriaan van Maanen, "The Relative Proper Motions of 162 Stars in

the Neighborhood of the Great Nebula in Orion," *Astronomical Journal* 27(1911):139–46.

34. Adriaan van Maanen, "The Photographic Determination of Stellar Parallaxes with the 60-inch Reflector," *Mount Wilson Contributions* 111(1915):1–33.
35. van Maanen, "Internal Motion in . . . Messier 33," 1–5.
36. van Maanen, "Internal Motion in . . . Messier 33," 1–5; R. Berendzen and R. Hart, "Adriaan van Maanen's Influence on the Island Universe Theory," *Journal for the History of Astronomy* 4(1973):46–56, 73–98. They cite a letter from van Maanen to Hale, 31 October 1916, in Daniel J. Kevles, ed., *George Ellery Hale Papers, 1882–1937* (Washington, D.C.: Carnegie Institution of Washington, 1967; Pasadena: California Institute of Technology, 1968), microfilm edition.
37. van Maanen, "Internal Motion in . . . Messier 33," 1–5.
38. Adriaan van Maanen, "Investigations on Proper Motion. Fourth Paper: Internal Motion in the Spiral Nebula Messier 51," *Astrophysical Journal* 54(1921):237–45.
39. Adriaan van Maanen, "Investigations on Proper Motion. Fifth Paper: Internal Motion in the Spiral Nebula Messier 81," *Astrophysical Journal* 54(1921):347–56; "Internal Motion in Four Spiral Nebulae," *Publications of the Astronomical Society of the Pacific* 33(1921):200–202.
40. Adriaan van Maanen to Harlow Shapley, 23 May 1921; van Maanen to Shapley, 1 June 1921; Shapley to van Maanen, 8 June 1921; Shapley to van Maanen, 21 July 1921, box 23b, Harlow Shapley Personal Papers, Harvard University Archives.
41. Adriaan van Maanen, "Investigations on Proper Motion. Seventh Paper . . . N.G.C. 2403," *Astrophysical Journal* 56(1922):200–7; ". . . Eighth Paper . . . M94 = N.G.C. 4736," ibid., 208–16; ". . . Ninth Paper . . . Messier 63, N.G.C. 5055," ibid., 57(1923):49–56.
42. van Maanen, ". . . Tenth Paper . . . Messier 33, N.G.C. 598," ibid., 24–278.
43. van Maanen, ". . . Eleventh Paper. . . . The Proper Motion of Messier 13 and Its Internal Motion," ibid., 61(1925):130–36.
44. van Maanen, ". . . Tenth Paper . . . Messier 33 . . .," 24–278.
45. Berendzen and Hart, "Adriaan van Maanen's Influence . . .," 46–56, 73–98; they cite a letter from van Maanen to Hale, 29 April 1916, *Hale Papers*.
46. Edwin Hubble, undated, three-page typed note in box 4, Hubble Collection. See also Hetherington, "Hubble's Examination of Internal Motions in Spiral Nebulae," 392–418.
47. van Maanen, ". . . Tenth Paper . . . Messier 33 . . .," 24–278.

48. Knut Lundmark, "The Spiral Nebula Messier 33," *Publications of the Astronomical Society of the Pacific* 33(1921):324-27.
49. Harlow Shapley to Knut Lundmark, 27 December 1921, file UAV 630.22, Director's Files, Harvard College Observatory, Harvard University Archives. See Norriss S. Hetherington, "New Source Material on Shapley, van Maanen, and Lundmark," *Journal for the History of Astronomy* 7(1976):73-74. Clark Elliott brought to my attention the existence of two different files of Shapley's correspondence.
50. Lundmark to Shapley, 2 January 1922, Director's Files, Harvard College Observatory.
51. Lundmark to Shapley, 3 January 1922, ibid.
52. Shapley to van Maanen, 14 January 1922, ibid.
53. van Maanen to Shapley, 9 June 1922, ibid.
54. Knut Lundmark, "On the Motions of Spirals," *Publications of the Astronomical Society of the Pacific* 34(1922):108-15.
55. Harlow Shapley to Adriaan van Maanen, 19 June 1922, Director's Files, Harvard College Observatory.
56. Shapley to Lundmark, 15 July 1922, ibid.
57. Shapley to van Maanen, 28 September 1922, ibid.
58. van Maanen to Shapley, 21 October 1922, ibid.
59. Lundmark to Shapley, 25 April 1923, ibid.
60. Shapley to van Maanen, 16 May 1923, ibid.
61. van Maanen, ". . . Tenth Paper . . . Messier 33 . . .," 24-278.
62. Hubble, three-page typed note, box 4, Hubble Collection.
63. Lundmark, "On the Motions of Spirals," 108-15.
64. Edwin Hubble, "Notes for a talk on the Application of the Cepheid Criterion to Spiral Nebulae," undated, handwritten, seventeen-page manuscript, box 4, Hubble Collection.
65. Hubble, three-page typed note, box 4, Hubble Collection.
66. Edwin Hubble, "Messier 51," undated, handwritten five-page paper, Hubble Collection.
67. Adriaan van Maanen to Harlow Shapley, 5 June 1921, Harlow Shapley Personal Papers.
68. Shapley to van Maanen, 8 September 1921, ibid.
69. Berendzen and Hart, "Adriaan van Maanen's Influence," 45-56, 73-98.
70. J. Jeans, "Internal Motion in Spiral Nebulae," *The Observatory* 40(1917):60-61. See also Jeans, *Problems of Cosmogony and Stellar Dynamics* (Cambridge: Cambridge Univ. Press, 1919), 203-19.
71. A. S. Eddington, "The Motions of Spiral Nebulae," *Monthly Notices of the Royal Astronomical Society* 77(1917):375-77.
72. "Meeting of the Royal Astronomical Society," *The Observatory* 40(1917):351-61.
73. J. H. Jeans, "Internal Motions in Spiral Nebulae," *Monthly Notices of the Royal Astronomical Society* 84(1923):60-76.

74. Ernest W. Brown, "Gravitational Forces in Spiral Nebulae," *Astrophysical Journal* 61(1925):97–113.
75. J. H. Jeans, "Note on the Distances and Structure of the Spiral Nebulae," *Monthly Notices of the Royal Astronomical Society* 85(1925):531–34; Ernest W. Brown, "Gravitational Motion in a Spiral Nebula," *The Observatory* 51(1928), 277–86.
76. Frederick H. Seares, "Adriaan van Maanen, 1884–1946," *Publications of the Astronomical Society of the Pacific* 58(1946):89–103.
77. Kreiken, "On the Differential Measurement of Proper Motion," 255–60.
78. W. M. Smart, "The Motions of Spiral Nebulae," *Monthly Notices of the Royal Astronomical Society* 84(1924):333–53.
79. Harlow Shapley, *Through Rugged Ways to the Stars* (New York: Charles Scribner's, 1968), 80.
80. Harlow Shapley to Henry Norris Russell, 3 September 1917, Harlow Shapley Personal Papers.
81. Shapley to van Maanen, 8 June 1921, ibid.
82. Walter S. Baade, *The Evaluation of Galaxies*, 27–28.
83. Berendzen and Hart, "Adriaan van Maanen's Influence," 46–56, 73–98; they cite letters from H. N. Russell to Harlow Shapley, 8 November 1917, and Russell to Adriaan van Maanen, 5 October 1920, Princeton University Archives.
84. Harlow Shapley to Adriaan van Maanen, 8 September 1921, and Shapley to van Maanen, 23 May 1921, Harlow Shapley Personal Papers.
85. Peter Doig, "The Spiral Nebulae," *Journal of the British Astronomical Association* 35(1925):99–105.
86. J. H. Reynolds, "Nebulae," *Monthly Notices of the Royal Astronomical Society* 84(1924):283–86.
87. Edwin Hubble to Harlow Shapley, 19 February 1924, Director's Files, Harvard College Observatory; an incomplete copy of the letter is in box 7, envelope 16, Hubble Collection.
88. Cecilia Payne-Gaposchkin with Katherine Haramundanis, ed., *An Autobiography and Other Recollections* (Cambridge: Cambridge Univ. Press, 1984), 209–10.
89. Harlow Shapley to Edwin Hubble, 27 February 1924, Director's Files, Harvard College Observatory.
90. Hubble to Shapley, 25 August 1924, ibid.; incomplete copy in Hubble Collection.
91. Shapley to Hubble, 5 September 1924, Director's Files, Harvard College Observatory.
92. Richard Berendzen and Michael Hoskin, "Hubble's Announcement of Cepheids in Spiral Nebulae," *Astronomical Society of the Pacific Leaflets* 504(1971):1–15; they quote a letter from Hubble to Henry Norris Russell, 19 February 1925. Documents relating to Hubble's

paper for the American Astronomical Society are preserved in the files of that society, which are deposited at the Center for History and Philosophy of Physics at the American Institute of Physics, New York City. Copies of the documents are in the Edwin Hubble Collection.

93. Edwin Hubble, "Notes for a talk on the Application of the Cepheid Criterion to Spiral Nebulae," seventeen-page undated manuscript, box 4, Hubble Collection.

94. Edwin Hubble, three pages, untitled, undated typescript discussing Adriaan van Maanen's claim of corroborative evidence, Hubble Collection. The 1923 van Maanen and 1925 Lundmark papers are: Adriaan van Maanen, "Investigations on Proper Motion. Tenth Paper: Internal Motion in the Spiral Nebula Messier 33, N.G.C. 598," *Astophysical Journal* 57(1923):264–78; Knut Lundmark, "The Motions and the Distances of Spiral Nebulae," *Monthly Notices of the Royal Astronomical Society* 85(1925):865–94. See also Lunkmark, "Internal Motions of Messier 33," *Astrophysical Journal* 63(1926):67–71.

95. Edwin Hubble, one-page undated manuscript on the manner in which stellar images build up on a photographic plate, Hubble Collection. Hubble's thoughts may have been stimulated by an article in the journal in which the manuscript page was found: F. G. Pease, "Notes on the Atmospheric Effects Observed with the 100-Inch Telescope," *Publications of the Astronomical Society of the Pacific* 36(1924):191–98.

96. Harlow Shapley to Edwin Hubble, 13 February 1925, and Hubble to Shapley, 11 March 1925, Director's File, Harvard College Observatory.

97. Berendzen and Hoskin, "Hubble's Announcement of Cepheids in Spiral Nebulae," 1–15; they quote a letter from Edwin Hubble to Joel Stebbins, 6 March 1925.

98. Grace Burke Hubble to Michael Hoskin, 7 March 1968, box 7, Hubble Collection.

99. Grace Hubble, undated note, box 5, Hubble Collection.

100. Richard Cullen Hart, "Adriaan van Maanen's Influence on the Island Universe Theory" (Ph.D. diss., Boston University, 1973), 167. There is, however, no documentary evidence of Adams's possible role and Hart's source has chosen not to repeat the suggestion (letter from H. W. Babcock to the author, 15 April 1975).

101. George E. Hale, Walter S. Adams, and Frederick H. Seares, "Mount Wilson Observatory," *Carnegie Institution of Washington Yearbook* 30(1931):171–221, especially 199–200, "Proper Motions Measured in Spiral Nebulae." See also Michael Anthony Hoskin, "Edwin Hubble and the Existence of External Galaxies," *Actes XIIᵉ Congrès International d'Histoire des Sciences* 5(1968):49–53.

102. Edwin Hubble, "Measures of M81 for Internal Motion," twelve-page

undated typescript, box 4, Hubble Collection. See also "Messier 51," five-page undated manuscript, and "Measures of M33 for Rotation," six-page undated typescript, box 4, Hubble Collection.

103. Edwin Hubble, three-page untitled undated manuscript discussing a possible magnitude error, box 4, Hubble Collection.

104. Norriss S. Hetherington, "Edwin Hubble on Adriaan van Maanen's Internal Motions in Spiral Nebulae," *Isis*, 65(1974):390–93; Hart, "Adriaan van Maanen's Influence on the Island Universe Theory," 167.

105. Edwin Hubble, "Internal Motions of Spiral Nebulae," 33-page undated typescript, earlier version, and seven accompanying tables, box 4, Hubble Collection. See also Hubble, "Angular Rotations of Spiral Nebulae," six-page undated typescript, box 4, Hubble Collection.

106. Robert W. Smith, *The Expanding Universe: Astronomy's 'Great Debate' 1900–1931* (Cambridge: Cambridge Univ. Press, 1982), 136.

107. Frederick Seares to George Ellery Hale, 24 January 1935, in Hale Papers, roll 33, frames 118–19.

108. Seares to Hale, 24 January 1935, in Hale Papers, roll 33, frame 120.

109. Ibid., 25 January 1935, frame 122.

110. Edwin Hubble, "Angular Rotations of Spiral Nebulae," *Astrophysical Journal* 81(1935):334–35.

111. Adriaan van Maanen, "Internal Motions in Spiral Nebulae," *Astrophysical Journal* 81(1935):336–37.

9. OBJECTIVITY QUESTIONED

1. Sir Arthur Eddington, *The Philosophy of Physical Sciences* (New York: Macmillan, 1939), 9, 16–19, 131.

2. Thomas Kuhn, *The Structure of Scientific Revolutions,* 2d ed. (Chicago: Univ. of Chicago Press, 1970).

3. Lewis M. Branscomb, "Integrity in Science," *American Scientist* 73(1985):421–23.

4. R. T. Birge, "A Survey of the Systematic Evaluation of the Universal Physical Constants," *Supplemento al vol. vi, serie x del Nuovo Cimento* 1(1957):39–67.

5. E. Richard Cohen and Jesse W. M. DuMond, "Our Knowledge of the Fundamental Constants of Physics and Chemistry in 1965," *Reviews of Modern Physics* 37(1965):537–94.

6. Robert Rosenthal, "Biasing Effects of Experimenters," *et cetra* 34(1977):253–64; "How Often Are Our Numbers Wrong," *American Psychologist* 33(1978):1005–8; *Experimenter Effects in Behavioral Research* (New York: Irvington, 1976); W. K. Estes, "Human Behavior in Mathematical Perspective," *American Scientist* 63(1975):649–55.

7. Rosenthal, "Biasing Effects"; John L. Kennedy and Howard F. Uphoff, "Experiments on the Nature of Extra-Sensory Perception. III. The Recording of Error Criticism of Extra-Chance Scores," *The Journal of Parapsychology* 3(1939):226–45.
8. Barbara J. Culliton, "The Sloan-Kettering Affair: A Story without a Hero," *Science* 184(1974):644–50; "The Sloan-Kettering Affair (II): An Uneasy Resolution," *Science* 184(1979):1154–57; L. S. Hearnshaw, *Cyril Burt: Psychologist* (Ithaca: Cornell Univ. Press, 1979); William Broad and Nicholas Wade, *Betrayers of the Truth* (New York: Simon and Schuster, 1982).
9. W. Baade, *Evolution of Galaxies* (Cambridge: Harvard Univ. Press, 1963), 28–30.
10. Richard Cullen Hart, "Adriaan van Maanen's Influence on the Island University Theory" (Ph.D. diss., Boston University, 1973), abstract.
11. Ibid., 220.
12. Michael J. Mahoney, "Psychology of the Scientist: An Evaluative Review," *Social Studies of Science* 9(1979):349–75.
13. Langdon Gilkey, "The Structure of Academic Revolutions," in *The Nature of Scientific Discovery*, ed. Owen Gingerich (Washington, D.C.: Smithsonian Press, 1975), 538–46.

Bibliographic Note

STUDIES of instances in which scientists have found what they expected to find even when it wasn't there to find have multiplied in recent years. The most important paper, however, remains H. Bondi's "Fact and Inference in Theory and in Observation," *Vistas in Astronomy* 1(1955):155-62. For a more recent review, see Norriss S. Hetherington, "Just How Objective Is Science?" *Nature* 306(1983):727-30.

Psychological studies of the presence of personal bias in science include papers by Robert Rosenthal, "How Often Are Our Numbers Wrong?" *American Psychologist* 33(1978):1005-8; by W. K. Estes, "Human Behavior in Mathematical Perspective," *American Scientist* 63(1975):649-55; and by John L. Kennedy and Howard F. Uphoff, "Experiments on the Nature of Extra-Sensory Perception. III. The Recording Error Criticism of Extra-Chance Scores," *The Journal of Parapsychology* 3(1939):226-45.

Early instances of spurious observational reports correlated with expectations have been studied by: Terrie F. Bloom, "Borrowed Perceptions: Harriot's Maps of the Moon," *Journal for the History of Astronomy* 9(1978):117-22; M. E. W. Williams, "Flamsteed's Alleged Measurement of Annual Parallax for the Pole Star," ibid. 10(1979):102-16; R. H. Austin, "Uranus Observed," *British Journal for the History of Science* 3(1967):275-84; Hetherington, "Neptune's Supposed Ring," *Journal of the British Astronomical Society* 90(1979):20-29; Robert W. Smith and Richard Baum, "William Lassell and the Ring of Neptune: A Case Study in Instrumental Failure," *Journal for the History of Astronomy* 15(1984):1-17; and Baum, *The Planets: Some Myths and Realities* (Newton Abbot: David and Charles, 1973).

Modern instances of failed objectivity in physics have been noted by George Magyar, "Pseudo-Effects in Experimental Physics:

Some Notes for Case Studies," *Social Studies of Science* 7(1977):241–67. One case was examined in detail by Mary Jo Nye, "N-rays: An Episode in the History and Psychology of Science," *Historical Studies in the Physical Sciences* 11(1980):125–56.

Adriaan van Maanen's twentieth-century observations of a general solar magnetic field were examined by Jan Stenflo, "Hale's Attempts to Determine the Sun's General Magnetic Field," *Solar Physics* 14(1970):263–73. See also Hetherington, "Adriaan van Maanen's Measurements of Solar Spectra for a General Magnetic Field," *Quarterly Journal of the Royal Astronomical Society* 16(1975):235–44. For another twentieth-century instance of failed objectivity, see Hetherington, "Sirius B and the Gravitational Redshift: An Historical Review," ibid. 21(1980):246–52.

The most thoroughly studied instance of a scientist finding what he expected to see when it wasn't there to see is van Maanen's alleged rotation of spiral nebulae. With the coming of widespread use of the telephone, which leaves few tracks, historians may never again be able to document in such detail a controversy within science.

Even in this case, historians are hindered by the presence of the two chief protagonists, Adriaan van Maanen and Edwin Hubble, at the same observatory. They scarcely needed to resort to written communication—if their personal relationship permitted any communication. Their letters to others, however, are invaluable, especially letters to Harlow Shapley at Harvard. These letters are preserved in the Harlow Shapley Papers and in the Files of the Director of the Harvard College Observatory; both sets of papers are held at the Harvard Library. Other Hubble letters are in the W. de Sitter papers at the Leiden Observatory and in the Lick Observatory Archives, now preserved at the library of the Santa Cruz Campus of the University of California.

The other major unpublished source of information on the controversy is in Hubble's unpublished manuscripts. They are on deposit at the Henry E. Huntington Library, of which Hubble was a trustee.

Other important depositories of relevant correspondence between astronomers, though not necessarily letters to or from Hubble or van Maanen, include the Lowell Observatory, whose files through 1916 are available on microfilm; the H. D. Curtis papers at the University of Michigan's Bently Library; the Henry Norris Russell papers at the Princeton Observatory; and the files of the Allegheny Observatory, now in the Archives of the Industrial Society at the University of Pittsburgh.

The Mount Wilson Observatory Archives contain at least one draft of a paper by van Maanen. When they are eventually opened, the archives may yield other documents of historical importance. Much of the correspondence to and from Mount Wilson astronomers is already available at the other end, as in the case of van Maanen's correspondence with Shapley, preserved at Harvard. The correspondence of George Ellery Hale, the director of the Mount Wilson Observatory, is at the California Institute of Technology and is available on microfilm.

Published scientific papers are to be found in a large number of journals, most often but not always in the major astronomical journals: *Astrophysical Journal, Monthly Notices of the Royal Astronomical Society,* and *Publications of the Astronomical Society of the Pacific.* Indexes to the vast astronomical literature are: J. C. Houzeau and L. Lancaster, *Bibliographie générale de l'astronomie jusqu'en 1880,* vol. 1 (parts 1 and 2) and vol. 2; *Bibliography of Astronomy, 1881–1898* (Tylers Green, Penn, High, Wycombe, Bucks, England: University Microfilms); *Astronomischer Jahresberichte,* vol. 1, 1899 through vol. 69, 1968; and *Astronomy and Astrophysics, A Bibliographic Guide,* beginning with vol. 1, 1969. For a review of bibliographic sources in the history of astronomy, see the *Journal for the History of Astronomy* 2(1971):50, 53–54, and 73.

Historical studies are of recent vintage. J. D. Fernie in "The Historical Quest for the Nature of the Spiral Nebulae," *Publications of the Astronomical Society of the Pacific* 83(1970):1189–1230, noted that a magnitude error would not serve to explain van Maanen's detection of motions consistently in the same direction with respect to the orientation of the arms of the spiral nebulae. A review of the scientific literature relevant to van Maanen's measurements appeared two years later in Hetherington's "Adriaan van

Maanen and Internal Motions in Spiral Nebulae: A Historical Review," *Quarterly Journal of the Royal Astronomical Society* 13(1972):25–39.

The first examination of relevant manuscript sources appeared the next year in Richard Berendzen and Richard Hart's "Adriaan van Maanen's Influence on the Island Universe Theory," *Journal for the History of Astronomy* 4(1973):46–56 and 73–98, and in Hart's "Adriaan van Maanen's Influence on the Island Universe Theory" (Ph.D. diss., Boston University, 1973). This work is presented again in Berendzen, Hart, and Daniel Seeley, *Man Discovers the Galaxies* (New York: Neale Watson Academic Publications, 1976). The draft of van Maanen's first paper on motions in spiral nebulae is described by Hetherington in "Adriaan van Maanen on the Significance of Internal Motions in Spiral Nebulae," *Journal for the History of Astronomy* 5(1974):52–53. The division of Harlow Shapley's papers between two different files is reported by Hetherington in "Additional Shapley—van Maanen Correspondence," ibid. 7(1976):73–74. Possible interaction between van Maanen and other observers is discussed by Hetherington in "The Simultaneous 'Discovery' of Internal Motions in Spiral Nebulae," ibid. 6(1975):115–35.

Hubble's unpublished manuscripts are described by Hetherington in "Edwin Hubble's Examination of Internal Motions in Spiral Nebulae," *Quarterly Journal of the Royal Astronomical Society* 15(1974):392–418. The specific issue of Hubble's examination of a possible magnitude error is sketched by Hetherington in "Edwin Hubble on Adriaan van Maanen's Internal Motions in Spiral Nebulae," *Isis* 65(1974):390–93.

Hubble's determination of the extragalactic nature of the spiral nebulae is examined in Berendzen and Michael Hoskin, "Hubble's Announcement of Cepheids in Spiral Nebulae," *Astronomical Society of the Pacific Leaflet* 504(1971):1–15; and in Hoskin, "Edwin Hubble and the Existence of External Galaxies," *XIIe Congrés International d'Histoire des Sciences* 5(1968):49–53. On Cepheid variables see also Fernie, "The Period-Luminosity Relation: A Historical Review," *Publications of the Astronomical Society of the Pacific* 81(1969):707–31.

Hoskin's work taken as a whole presents a remarkably comprehensive history of modern stellar astronomy. In addition to his papers on Hubble and on Cepheids cited above, see his *William Herschel and the Construction of the Heavens* (London: Oldbourne, 1963); "Stellar Distances: Galileo's Method and its Subsequent History," *Indian Journal of History of Science* 1(1966):22–29; "Apparatus and Ideas in Mid-nineteenth-century Cosmology," *Vistas in Astronomy* 9(1967):79–85; introductions to Hoskin's editions of Thomas Wright, *An Original or New Hypothesis of the Universe* (London: Macdonald, 1971); *Clavis Coelestis* (London: Dawson, 1967), and *Second or Singular Thoughts Upon the Theory of the Universe* (London: Dawson, 1968); "The Cosmology of Thomas Wright of Durham," *Journal for the History of Astronomy* 1(1970):44–52; "Lambert and Herschel," ibid. 9(1978):140–42; "The English Background to the Cosmology of Wright and Herschel," in *Cosmology, History, and Theology,* ed. W. Yourgrau and A. Breck (New York: Plenum Press, 1977), 219–31; "The 'Great Debate': What Really Happened," *Journal for the History of Astronomy* 7(1976):169–82; "Ritchey, Curtis, and the Discovery of Novae in Spiral Nebulae," ibid. 47–53. Several of these essays are collected in Hoskin's *Stellar Astronomy* (Chalfont St Giles: Science History Publications, 1982). This volume also contains new material.

On astronomical photography, see Deborah Jean Warner (neé Mills), "The American Photographical Society and the Early History of Astronomical Photography in America," *Photographic Science and Engineering* 11(1967):342–47; and "George Willis Ritchey and the Development of Celestial Photography," *American Scientist* 54(1966):64–93. Astronomical photography and spectroscopy are discussed in Hetherington's "Observational Cosmology in the Twentieth Century," in *Human Implications of Scientific Advance,* ed. E. G. Forbes (Edinburgh: Edinburgh Univ. Press, 1978), 567–75.

Astronomical spectroscopy is discussed in Hetherington's "The Discovery and Early Application of the Velocity Distance Relationship" (Ph.D. diss., Indiana University, 1970); and "The Measurement of Radial Velocities of Spiral Nebulae," *Isis* 62(1971):309–13.

These studies use only printed sources. More recent work using manuscript material is that of Robert W. Smith, "The Origins of the Velocity-Distance Relation," *Journal for the History of Astronomy* 10(1979):133–65. See also Smith's *The Expanding Universe: Astronomy's 'Great Debate' 1900–1931* (Cambridge: Cambridge Univ. Press, 1982).

Index

Preconception, 7-8, 14, 20, 21, 27-
28, 29, 31, 51, 61, 70, 80-81,
85, 91, 110, 112, 117
Princeton University, 101
Purpose and Design, 3-4
Pygmalion experiment, 114

Radcliffe Observatory, 26
Raleigh, Sir Walter, 11
Random error, 113
Rational Determinism, 4
Reaves, Gibson, xii
Relativity theory
bending of light, 67, 68, 69
gravitational redshift, 67-72
lost values, 3
perihelion of Mercury, 67, 70
Resolution of nebulae, 83
Retrodiction, 68, 80
Richard II (Shakespeare), 122
Ritchey, George W., 85, 86, 90, 92,
94, 95, 105
Ronan, Colin, xii
Rosenthal, Robert, 157
Rosse, earl of, 39, 46, 83
Rotation of spiral nebulae, 75, 83-
110, 121
Royal Astronomical Society, ix-x, 100,
102, 105
Royal Observatory (Cape Town), 74
Royal Observatory (Greenwich), 21,
38, 59
Royal Society of London, 17-19, 27,
39, 70
Russell, Henry Norris, 8, 53, 61, 81,
97, 101, 120

Sarton, George, 4-5
Schaffer, Simon, xii, 27
Schiaparelli, Giovanni, 50
Schouten, W. J. A., 92, 95, 99

Science and human values, x-xi, 5
Science and Human Values
(Bronowski), x
Seares, Frederick, 107-9, 120
See, T., J., J., 66
Selection effect, 112
Shakespeare, 122
Shapley, Harlow, 84, 94-95, 96-98,
101-4, 120, 122
Shaw, Bernard, 3-4
Sirius, 66
Sirius B, 66-72, 73, 113, 121
Slipher, Vesto M., 58, 84, 89, 119
Smart, W. M., 101
Smith, Robert, xii, 40, 45, 46-47,
107, 134, 135, 157, 162
Sociology of science, 8
Socrates, ix
Speed of light, 112-13
Spiral nebulae, 75, 83-110
Stellar aberration, 19
Stellar parallax, 16-21
Stenflo, Jan, 80-81, 158
Stereocomparator, 76, 86, 88, 93, 101
Sun
eclipse expedition, 67
general solar magnetic field, 75-82,
117, 143
gravitational redshift, 67
plages (floculi), 75
rotation, 76
sunspots, 15-16, 75

Tacitus, ix
Telescope, 11, 16
Theology, 83, 125
Theories in science, 7
Thiessen, G., 79-80, 143
Times (London), 39, 40
To Mars Via the Moon (Wicks), 52
Too True to Be Good (Shaw), 3-4
Triton, 40